T0255639

Block Transceivers

OFDM and Beyond

Synthesis Lectures on Communications

Block Transceivers: OFDM and Beyond
Paulo S.R. Diniz, Wallace A. Martins, and Markus V.S. Lima
2012

Basic Simulation Models of Phase Tracking Devices Using MATLAB
William Tranter, Ratchaneekorn Thamvichai, and Tamal Bose
2010

Joint Source Channel Coding Using Arithmetic Codes
Dongsheng Bi, Michael W. Hoffman, and Khalid Sayood
2009

Fundamentals of Spread Spectrum Modulation
Rodger E. Ziemer
2007

Code Division Multiple Access(CDMA)
R. Michael Buehrer
2006

Game Theory for Wireless Engineers
Allen B. MacKenzie and Luiz A. DaSilva
2006

Block Transceivers: OFDM and Beyond

Paulo S.R. Diniz, Wallace A. Martins, and Markus V.S. Lima

ISBN: 978-3-031-00549-7 paperback
ISBN: 978-3-031-01677-6 ebook

DOI 10.1007/978-3-031-01677-6

A Publication in the Springer series
SYNTHESIS LECTURES ON COMMUNICATIONS

Lecture #7
Series Editor: William Tranter, *Virginia Tech*
Series ISSN
Synthesis Lectures on Communications
Print 1932-1244 Electronic 1932-1708

Block Transceivers

OFDM and Beyond

Paulo S.R. Diniz, Wallace A. Martins, and Markus V.S. Lima
Universidade Federal do Rio de Janeiro

SYNTHESIS LECTURES ON COMMUNICATIONS #7

ABSTRACT

The demand for data traffic over mobile communication networks has substantially increased during the last decade. As a result, these mobile broadband devices spend the available spectrum fiercely, requiring the search for new technologies. In transmissions where the channel presents a frequency-selective behavior, multicarrier modulation (MCM) schemes have proven to be more efficient, in terms of spectral usage, than conventional modulations and spread spectrum techniques.

The orthogonal frequency-division multiplexing (OFDM) is the most popular MCM method, since it not only increases spectral efficiency but also yields simple transceivers. All OFDM-based systems, including the single-carrier with frequency-division equalization (SC-FD), transmit redundancy in order to cope with the problem of interference among symbols. This book presents OFDM-inspired systems that are able to, at most, halve the amount of redundancy used by OFDM systems while keeping the computational complexity comparable. Such systems, herein called memoryless linear time-invariant (LTI) transceivers with reduced redundancy, require low-complexity arithmetical operations and fast algorithms. In addition, whenever the block transmitter and receiver have memory and/or are linear time-varying (LTV), it is possible to reduce the redundancy in the transmission even further, as also discussed in this book. For the transceivers with memory it is possible to eliminate the redundancy at the cost of making the channel equalization more difficult. Moreover, when time-varying block transceivers are also employed, then the amount of redundancy can be as low as a single symbol per block, regardless of the size of the channel memory.

With the techniques presented in the book it is possible to address what lies beyond the use of OFDM-related solutions in broadband transmissions.

KEYWORDS

block transceivers, multicarrier modulation (MCM), orthogonal frequency-division multiplexing (OFDM), reduced-redundancy transceivers, broadband digital communications

Contents

Preface

The widespread use of mobile devices with high processing capabilities, like smartphones and tablets, as well as the increasing number of users and growing demand for higher data rates are some of the main reasons why data traffic over mobile communication networks has increased so much during the last decade. As a result, these mobile broadband devices spend the available spectrum fiercely, requiring the search for new technologies.

From the physical-layer viewpoint, the first step to address this problem is to choose a modulation scheme which is more adequate for the type of channel through which the signal (wave) propagates. Indeed, the spectral efficiency of communications systems can significantly increase by properly choosing the modulation scheme. For example, in broadband transmissions in which the channel presents a frequency-selective behavior, multicarrier modulation (MCM) schemes have proven to be more efficient, in terms of spectral usage, than conventional modulations and spread spectrum techniques. The multicarrier transmission illuminates the physical channel utilizing several non-overlapping narrowband subchannels, where each subchannel appears to be flat, thus turning the equalization process simpler.

Among the existing MCM schemes, the orthogonal frequency-division multiplexing (OFDM) is the most notorious, since it not only increases spectral efficiency but also yields simple transceivers. OFDM is capable of eliminating the intersymbol interference (ISI) with very simple transmitter and receiver, by performing low-complexity computations such as insertion and removal of a prefix, and by using fast algorithms such as the fast Fourier transform. It is worth noting that ISI is one of the most harmful effects in broadband transmissions. Simple transceivers are attractive since they lead to lower latency and require less power consumption. Therefore, it is no surprise that OFDM has been adopted by many wired and wireless broadband communication technologies. For instance, the long-term evolution (LTE) is a wireless communication standard whose downlink connection is based on OFDM, whereas its uplink connection is based on the single-carrier with frequency-division equalization (SC-FD), which is similar to OFDM in many aspects and is composed of the same building blocks. The enhancements introduced by LTE physical layer are so significant that LTE achieves much higher data rates, as compared to 3rd generation (3G) systems, and is already being considered a 4th generation (4G) system.

At this moment, given the desirable features of OFDM and SC-FD schemes and their widespread use in both wired and wireless communication standards, one could ask the following questions: Is this the best we can do in terms of spectral efficiency? When LTE spectrum gets overloaded, what comes next? Can we further improve these schemes?

This book tries to provide some directions to address these questions. Both OFDM and SC-FD systems transmit redundancy (the prefix) in order to cope with the problem of ISI. Indeed, a portion of each data block is reserved for the prefix, whose size must be larger than the channel memory for OFDM and SC-FD systems. Thus, spectral efficiency may be increased if less redundancy could be used. This book presents OFDM-inspired systems that are able to, at most, halve the amount of redundancy used by OFDM systems while keeping the computational complexity comparable. Such systems, herein called memoryless linear time-invariant (LTI) transceivers with reduced redundancy, require low-complexity arithmetical operations and fast algorithms. In addition, whenever the block transmitter and receiver have memory and/or are linear time-varying (LTV), it is possible to reduce the redundancy in the transmission even further, as discussed in the last chapter of this book. For the transceivers with memory it is possible to eliminate the redundancy at the cost of making the channel equalization a more difficult task. Moreover, when time-varying block transceivers are also employed, then the amount of redundancy can be as low as a single symbol per block, regardless of the size of the channel memory. An example is the code-division multiple access (CDMA) system with long spreading codes which is always able to achieve ISI elimination as long as the system is not at full capacity, i.e., at least one spreading/de-spreading code is unused.

The approach followed in this book is to present both OFDM and SC-FD systems as particular cases of the so-called transmultiplexer (TMUX). In fact, these two systems belong to the category of memoryless TMUXes. Special attention is given to OFDM and its desirable properties, since OFDM is being employed in many standards. Then, the TMUX is used as the main framework to derive both LTI and LTV transceivers with reduced redundancy.

With the techniques presented in the book it is possible to address what lies beyond the use of OFDM-related solutions in broadband transmissions. In summary, this book presents solutions to reduce the redundancy in transmission aiming at increasing data throughput. However, it is worth mentioning that reducing redundancy might increase mean square error (MSE) and bit-error rate (BER), as well as turn the design of the block transceiver more challenging. The optimum solutions are environment-dependent and its proper sensing leads to much more efficient spectral usage.

ORGANIZATION OF THE BOOK

Chapter 1 aims at providing a big picture of digital and wireless communications. This chapter differs from the others in the sense that the material is presented in a pictorial manner, avoiding the mathematics whenever possible. The reasons for this choice are: (i) to provide a quick overview of the field without wasting time explaining concepts that are not central to this book, most of which are thoroughly explained in digital communication and wireless communication textbooks; and (ii) to concentrate on ideas rather than mathematics. Therefore, at the end of this chapter the reader should have recollected topics such as: digital modulation, channel encoder, cellular systems, multiple access methods, frequency-selective channels, multicarrier modulation schemes, and OFDM.

Chapter 2 briefly presents multirate signal processing fundamentals that are of major importance to fully understand the TMUX framework, which is employed throughout the rest of the book. In this chapter, it is shown that TMUXes are general structures that can be used to represent/model several communications systems. In particular, OFDM and SC-FD systems can be interpreted as memoryless TMUXes, whose implementations are based on memoryless block-based transceivers. This chapter also introduces some initial results related to what is beyond OFDM-based systems.

Chapter 3 introduces OFDM from its original analog conception to its actual discrete-time practical usage. The chapter starts with the analog OFDM, exploring the role played by the guard interval in maintaining the orthogonality among OFDM subcarriers, and explaining the choices for some parameters such as the OFDM symbol duration and distance between adjacent subcarriers. Then, the discrete-time OFDM is studied in connection with its analog version. In addition, many other topics related to OFDM are covered, such as the coded OFDM (C-OFDM), issues of OFDM transmissions like the peak-to-average power ratio (PAPR), discrete multitone (DMT) systems, and optimal power allocation.

Chapter 4 presents multicarrier and single-carrier memoryless LTI transceivers that use a reduced amount of redundancy, as compared to OFDM and SC-FD systems. That is, this chapter describes how transceivers with reduced redundancy can be implemented employing superfast algorithms based on the concepts of structured matrix representations. Thus, part of this chapter describes structured matrices and the displacement theory that allows the derivation of these superfast algorithms.

The focus of Chapter 5 is on the fundamental limits of some parameters related to LTV transceivers. In particular, we consider the memory of the multiple-input multiple-output (MIMO) receiver matrix and the number of transmitted redundant elements, which are inherent to finite impulse response (FIR) LTV transceivers satisfying the zero-forcing (ZF) constraint. In addition, it is shown that ZF equalizers cannot be achieved when no redundancy is used, and as alternative pure MMSE-based solutions are presented.

PREREQUISITES

We attempted to make this book as self contained as possible. Although basic knowledge of wireless communications, digital transmission, and multirate signal processing is highly desirable, it is not necessary since the first two chapters revisit the main concepts which are used throughout the book. Thus, the main prerequisites to follow this book are: digital communications, basic concepts of stochastic processes—involving expected values, means, and variances of random variables—and linear algebra—involving operations with vectors and matrices, ranks, determinants, null and range spaces. More advanced concepts, such as structured matrices and displacement theory, are explained.

Paulo S.R. Diniz, Wallace A. Martins, and Markus V.S. Lima
Rio de Janeiro
June 2012

Acknowledgments

The authors are grateful to Joel Claypool for kindly pushing us to finish this project. They are also thankful to Professors M.L.R. de Campos, E.A.B. da Silva, L.W.P. Biscainho, and S.L. Netto of UFRJ, and Professor T.N. Ferreira of UFF for their incentive and always being available to answer our questions. They would like to thank Professors R. Sampaio Neto of PUC-RJ and V.H. Nascimento of USP who influenced some parts of this book. Wallace thanks his colleagues at the Federal Center for Technological Education Celso Suckow da Fonseca (CEFET/RJ–UnED-NI), in particular at the Department of Control and Automation Industrial Engineering.

We also would like to thank our families for their patience and support during this challenging process of writing a book. Paulo would like to thank his parents, his wife Mariza, and his daughters Paula and Luiza for illuminating his life. Wallace thanks his fiancee Claudia and his parents, Renê and Perpétua. Markus thanks his parents, Luiz Álvaro and Aracy, and his girlfriend Bruna.

Paulo S.R. Diniz
Wallace A. Martins
Markus V.S. Lima
Rio de Janeiro, Brazil
June 2012

List of Abbreviations

2G	2nd Generation
3G	3rd Generation
3GPP	3rd Generation Partnership Project
4G	4th Generation
ADSL	Asymmetric Digital Subscriber Line
BER	Bit Error Rate
CDMA	Code-Division Multiple Access
CFO	Carrier-Frequency Offset
CP-OFDM	Cyclic-Prefix Orthogonal Frequency-Division Multiplexing
CP-SC-FD	Cyclic-Prefix Single-Carrier with Frequency-Division equalization
CSI	Channel-State Information
DAB	Digital Audio Broadcasting
DFT	Discrete Fourier Transform
DHT	Discrete Hartley Transform
DMT	Discrete MultiTone
DS-CDMA	Direct Sequence CDMA
DSP	Digital Signal Processing
ETSI	European Telecommunications Standards Institute
FDD	Frequency-Division Duplex
FDM	Frequency-Division Multiplexing
FDMA	Frequency-Division Multiple Access
FFT	Fast Fourier Transform
FH-CDMA	Frequency-Hopping CDMA
FIR	Finite Impulse Response
GSM	Global System for Mobile communications
IBI	InterBlock Interference
ICI	InterCarrier Interference
IDFT	Inverse Discrete Fourier Transform
IEEE	Institute of Electrical and Electronics Engineers
IIR	Infinite Impulse Response
ISI	InterSymbol Interference
LAN	Local Area Network
LTE	Long Term Evolution
LTI	Linear Time-Invariant

MA	Multiple Access
MAN	Metropolitan Area Network
Mbps	Megabits per second
MC-MRBT	MultiCarrier Minimum-Redundancy Block Transceiver
MC-RRBT	MultiCarrier Reduced-Redundancy Block Transceiver
MIMO	Multiple-Input Multiple-Output
MISO	Multiple-Input Single-Output
MMSE	Minimum Mean Square Error
MSC	Mobile Switching Center
MSE	Mean Square Error
MUI	MultiUser Interference
OFDM	Orthogonal Frequency-Division Multiplexing
OFDMA	Orthogonal Frequency-Division Multiple Access
OLA	Overlap-And-Add
PAM	Pulse-Amplitude Modulation
PAN	Personal Area Network
PAPR	Peak-to-Average Power Ratio
PSD	Power Spectrum Density
PSK	Phase-Shift Keying
QAM	Quadrature Amplitude Modulation
QPSK	Quadrature PSK
SC	Single Carrier
SC-FD	Single-Carrier with Frequency-Domain equalization
SC-FDMA	Single-Carrier Frequency-Division Multiple Access
SC-MRBT	Single-Carrier Minimum-Redundancy Block Transceiver
SC-RRBT	Single-Carrier Reduced-Redundancy Block Transceiver
SGSN	Serving GPRS Support Node
SIMO	Single-Input Multiple-Output
SISO	Single-Input Single-Output
SNR	Signal-to-Noise Ratio
SVD	Singular Value Decomposition
VDSL	Very high-speed Digital Subscriber Line
TDD	Time-Division Duplex
TDM	Time-Division Multiplexing
TDMA	Time-Division Multiple Access
TMUX	Transmultiplexer
UMTS	Universal Mobile Telecommunications System

WAN	Wide Area Network
Wi-Fi	Wireless Fidelity
WiMAX	Worldwide interoperability for Microwave ACCess
WLAN	Wireless Local Area Network
WPAN	Wireless Personal Area Network
WSS	Wide-Sense Stationary
xDSL	high-speed Digital Subscriber Line
ZF	Zero-Forcing
ZP	Zero-Padding
ZP-OFDM	Zero-Padding OFDM
ZP-OFDM-OLA	ZP-OFDM OverLap-and-Add
ZP-SC-FD	Zero-Padding SC-FD
ZP-SC-FD-OLA	ZP-SC-FD OverLap-and-Add
ZP-ZJ	Zero-Padding Zero-Jamming

List of Notations

Scalars	Lowercase letters, e.g., x
Vectors	Lowercase boldface letters, e.g., $\mathbf{x}$
Matrices	Uppercase boldface letters, e.g., $\mathbf{X}$
$\triangleq$	Definition
$\mathbb{N}$	Set of natural numbers, which is defined as $\mathbb{N} \triangleq \{1, 2, 3, \ldots\}$
$\mathbb{Z}$	Set of integer numbers
$\mathbb{R}$	Set of real numbers
$\mathbb{C}$	Set of complex numbers
t	Real-valued variable representing continuous time
n	Integer number representing discrete time
j	Imaginary unit, $\mathrm{j}^2 = -1$
ω	Angular frequency
$\delta(t)$	Dirac impulse
$\delta[n]$	Kronecker delta
$\mathbf{W}_M$	Unitary DFT matrix of size $M \times M$
$\mathbf{I}_M$	Identity matrix of size $M \times M$
$\mathbf{e}_m$	Canonical vector, e.g., $\mathbf{e}_0 = [1\ 0\ \ldots\ 0]^T$
$(\cdot)^T$	Transpose of matrix $(\cdot)$
$(\cdot)^H$	Hermitian (conjugate) transpose of matrix $(\cdot)$
$\mathbf{0}_{M \times N}$	$M \times N$ matrix with all entries equal to 0
$E[(\cdot)]$	Expected value of $(\cdot)$
$\mathcal{Z}\{(\cdot)\}$	$\mathcal{Z}$-transform applied to $(\cdot)$
$\mathcal{Z}^{-1}\{(\cdot)\}$	Inverse $\mathcal{Z}$-transform applied to $(\cdot)$
$\mathrm{tr}\{(\cdot)\}$	Trace of matrix $(\cdot)$
$\mathrm{rank}\{(\cdot)\}$	Rank of matrix $(\cdot)$
$\mathrm{diag}\{(\cdot)\}$	Diagonal matrix whose entries in its diagonal are $(\cdot)$
$\ker\{(\cdot)\}$	Kernel (Null space) of matrix $(\cdot)$
$\mathcal{R}\{(\cdot)\}$	Range (Column space) of matrix $(\cdot)$
$\mathcal{F}\{(\cdot)\}$	Fourier transform of $(\cdot)$
$[(\cdot)]_{\downarrow N}$	Decimation operator by N
$[(\cdot)]_{\uparrow N}$	Interpolation operator by N
$[(\cdot)]_{ml}$	Entry of matrix $(\cdot)$ in the mth row and lth column
$\mathrm{DFT}\{(\cdot)\}$	Discrete Fourier transform of sequence $(\cdot)$

List of Notations

CHAPTER 1
The Big Picture

1.1 INTRODUCTION

In communications the ultimate goal is the transport of as much information as possible through a propagation medium. Currently, most transmissions are performed by propagating electromagnetic energy through the air or wired channels. The majority of our current communications systems transmit digital data to benefit from the widely available digital technology which has become relatively cheap. The digital technology is also reliable, amenable to error detection and correction, and reproducible. All these features reduce the cost of transmission and make available new services to the end users.

The wired channels involve a physical connection between fixed communication terminals, usually consisting of guided electromagnetic channels such as twisted-pair wirelines and coaxial cables. In addition, optical fibers are becoming increasingly popular due to their channel bandwidth. Indeed, the bandwidth of optical fiber is in general some orders of magnitude larger than in coaxial cables. Even though there is a trend for replacing wireline by optical-fiber channels, the low costs and improvements in modem designs have extended the lifetime of many wireline connections which were already deployed.

In wireless communications, the channel is the medium through which the electromagnetic energy propagates. Examples of such mediums are the air and the water. In this book, the focus is on wireless communications through the air, which we will refer to only as *wireless communications*.[1] In such communications systems, the electromagnetic energy is radiated to the propagation medium via an antenna.

Unlike wired transmissions, wireless transmissions require the use of radio spectrum, which in turn should be carefully managed by government regulators. The current trend of increasing the demand for radio transmissions shows no sign of settling. The amount of wireless data services is more than doubling each year leading to spectrum shortage as a sure event in the years to come. As a consequence, all efforts to increase the efficiency of spectrum usage are highly justifiable at this point. In response this book addresses the issue of how to increase the spectral efficiency of radio links by properly designing the transceivers, especially for multicarrier systems, which include the popular orthogonal frequency-division multiplexing (OFDM) as a special case.

This chapter starts with a brief description of digital communications systems in Section 1.2. Section 1.3 motivates the study of OFDM by giving examples of several communications standards that use it, with emphasis to the WiMAX and LTE systems. The next two sections contain key

[1]Communications through the water are usually referred to as underwater communications.

concepts of mobile and multiuser communications. Indeed, Section 1.4 introduces the cellular-division paradigm, while Section 1.5 briefly describes the most used multiple access schemes and also explains how multiple access can be performed in OFDM-based systems, leading to the so-called OFDMA (orthogonal frequency-division multiple access). After that, Section 1.6 describes duplex methods. Then, Section 1.7 introduces the main problems present in wireless communications systems, emphasizing the multipath effect. In Section 1.8, a central topic of this book, namely block transceivers, is introduced. In this section, the use of a guard time between blocks is exemplified as a naive solution to avoid interblock interference. Section 1.9 presents the fundamental idea of multicarrier systems, which is the division of the channel spectrum in narrowband and approximately flat fading subchannels, that allows simple equalization schemes at the receiver, as it will be discussed in the following chapters. Section 1.10 shows for the first time the simple, yet powerful, mathematical model for representing OFDM as a MIMO (multiple-input multiple-output) system. The target is to illustrate that the low complexity of the OFDM is due to parameter decoupling, which is achieved by judicious design of both the precoder (transmitter) and postcoder (receiver). Finally, in Sections 1.11 and 1.12 we briefly discuss multiple antenna systems and the interference issue, respectively.

1.2 DIGITAL COMMUNICATIONS SYSTEMS

Using a simplified building-block representation, the main elements of a digital communication system are depicted in Figure 1.1 (see [4, 7, 22, 26, 47, 57, 65, 66, 75] for further details).

Firstly, typical transmitted *input signals*, like data, speech, audio, image, and video, are compressed at the *source encoder*. By exploring the redundancy[2] of the input signal, the source encoder is capable of representing it in a more compact form, i.e., using less bits. Depending on the input signal nature, the compression process can be either lossless, which means that the original signal can be exactly recovered from the compressed signal, or lossy, which does not allow exactly restoration of the input signal, but yields a much higher compression rate. Lossy compression schemes use perceptive criteria to discard information that is not perceived by the end users. For instance, an audio signal at the receiver end should keep the perceptive quality as close as possible to the original input signal.

The next building block is the *channel encoder* whose primary task is to protect the compressed input information against the physical-channel impairments. The channel encoder allows for error detection and correction by adding some bits to the compressed signal. This building block acts as a wrap to encase and secure the information to be transported [38, 90].

Once our package of information is ready, it must be represented in a proper format to cross the *channel*. This task is performed by the *digital modulator* which is responsible for mapping bits into waveforms. These waveforms are mathematically represented by complex numbers and are usually called *symbols*. After crossing the channel the received waveform (symbol) is usually a distorted

[2] Here, the term redundancy means predictability, which is a common characteristic of natural signals, and is related to the concept of entropy [14]. In the rest of the book the term redundancy is related to the prefix or suffix used in block transmissions, a topic that will be covered in detail in the following chapters.

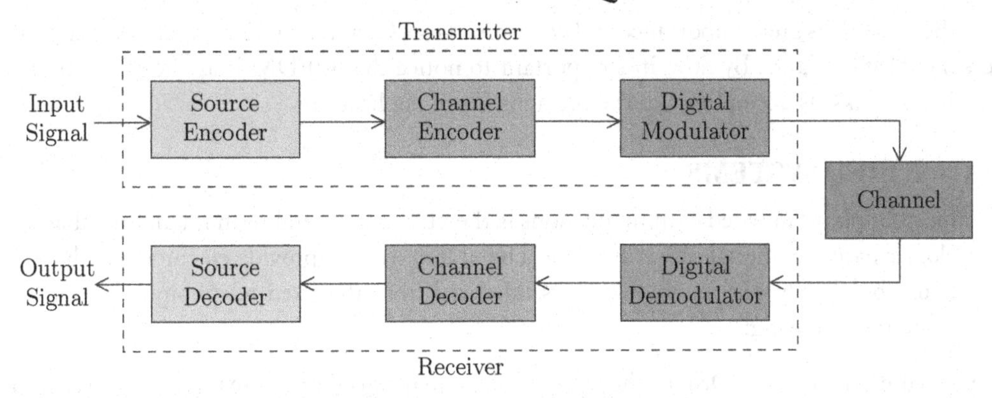

Figure 1.1: Simplified representation of digital communication systems.

version of the transmitted waveform since the latter suffers attenuation and other wave propagation effects, and interferences caused by environmental noise and by other signals being transmitted through the same channel.

At the receiver end, all the strategies utilized at the transmitter to improve the channel usage are undone. The received waveform is converted to bits by the *digital demodulator*, the *channel decoder* is responsible for correcting some bits that were erroneously detected, and then the *source decoder* undoes the compression process, which is not a perfect reversion process in cases of using lossy compression schemes (as explained before), generating an *output signal* as close as possible to the input signal.

The next section addresses the OFDM, a transmission scheme that enables efficient transmission of symbols (waveforms) through the channel.

1.3 ORTHOGONAL FREQUENCY-DIVISION MULTIPLEXING

Orthogonal frequency-division multiplexing (OFDM) is a transmission technique that is currently used in a number of practical systems, such as:

- Digital radio broadcasting or digital audio broadcasting (DAB);
- Wireless local area network (WLAN);
- Wireless broadband links;
- High-speed digital subscriber line (xDSL);
- 4th generation (4G) cellular communications;
- Digital broadcasting TV.

Because of its great importance and widespread use, Chapter 3 is dedicated to explain OFDM and some of its variants. By now, it is important to notice that OFDM is applicable to both wired and wireless links, as exemplified in the previous list of applications.

1.3.1 WIRED SYSTEMS

An important player in wired communications is the xDSL, a general term for all broadband access technologies based on digital subscriber line. The xDSL systems provide customers with high data rates using the already existing copper pairs inherited from the fixed telephony. The main xDSL systems are the following:

- Asymmetric DSL (ADSL) – the asymmetric term means that upstream (i.e., from costumer to network) and downstream (i.e., from network to costumer) data rates are different. Nowadays, evolved versions of the ADSL, such as the ADSL2+M, can achieve upstream rates up to 3.5 Mbps (megabits per second) and downstream rates up to 24 Mbps.

- Very high-speed DSL (VDSL) – the first VDSL systems, known as VDSL1, provided data rates higher than the ADSL. The drawback of VDSL1 is that its data rate decreases too fast as the distance from the subscriber premises to the network increases, limiting its usage to short local loops. Nowadays, VDSL2 systems can achieve upstream rates up to 10 Mbps and downstream rates up to 50 Mbps, and perform quite similar to the ADSL2+ when transmitting over long distances, unlike VDSL1.

Both ADSL and VDSL use discrete multitone (DMT), which is essentially a sophistication of OFDM.

1.3.2 WIRELESS SYSTEMS AND NETWORKS

A set of complementary wireless standards are available. These standards are mainly divided according to their coverage area and their main classes are:[3]

- Personal area network (PAN);

- Local area network (LAN);

- Metropolitan area network (MAN);

- Wide area network (WAN).

Figure 1.2 depicts these main classes according to their coverage area, and also shows examples of wireless standards. Note that some technologies may appear in more than one class of wireless networks, since they may receive upgrades that extend their original coverage area. For instance,

[3] Some of these classes also exist for wired systems. Indeed, PAN and LAN can be used for both wired and wireless systems. In cases when one wants to refer just to the wireless part of these systems, the terms wireless PAN (WPAN) and wireless LAN (WLAN) can be used.

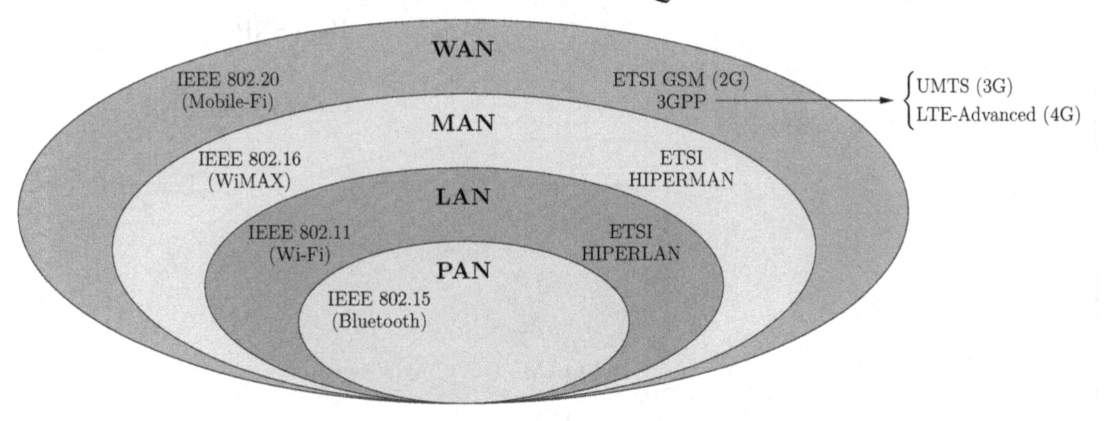

Figure 1.2: Wireless networks and corresponding technologies.

the WiMAX (worldwide interoperability for microwave access) is usually considered an example of MAN, but it can also be seen as a WAN system.

Many wireless network standards use OFDM in the air interface. Two examples are the LTE (long-term evolution) and WiMAX standards. LTE was designed to fully replace the 3rd generation (3G) networks for mobile communications. The WiMAX, although originally conceived to provide wireless broadband services to homes, has been upgraded to be employed by mobile phones as access method in recent years, competing with LTE.

(A) WiMAX

The main target of IEEE 802.16 standard, known as WiMAX, is to deliver wireless high-speed Internet access over longer distances than the ones supported by the IEEE 802.11 standard, commonly known as Wi-Fi (wireless fidelity). Indeed, WiMAX provides a MAN with wireless broadband service in an area of 50 km (about 30 miles) of radius. The WiMAX is guided by an association called WiMAX Forum and its data rates can reach up to 40 Mbps for low mobility access and up to 15 Mbps for mobile access. The capabilities and coverage area of WiMAX systems, especially due to the amendment e of IEEE 802.16 standard (IEEE 802.16e, also called Mobile WiMAX), increased in such a way that the IEEE 802.20 standard was put to hibernation.

WiMAX employs an adaptive modulation scheme as illustrated in Figure 1.3 in which the digital modulator is adjusted according to the *signal-to-noise ratio* (SNR). Through low-SNR channels, usually consisting of channels where the user is far from the base station, a sparser modulation scheme, the *quadrature phase-shift keying* (QPSK), is used. On the other hand, users near the base station are likely to have high-SNR channels and, therefore, transmissions with higher-order modulation schemes, such as the 16- or 64-*quadrature amplitude modulation*

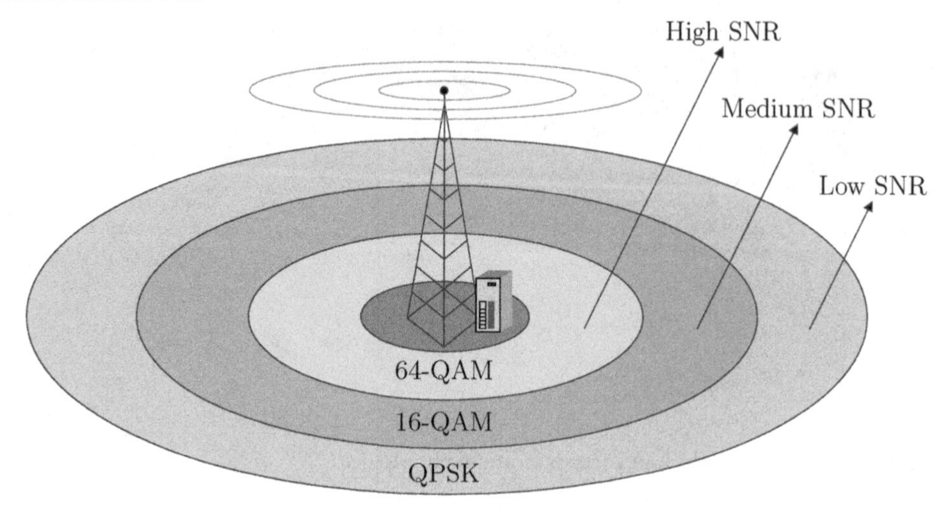

Figure 1.3: Adaptive modulation in WiMAX.

(QAM), are allowed. In addition, the multiple access scheme used in WiMAX is the orthogonal frequency division multiple access (OFDMA).

Other goals of WiMAX are:

- Broadband on demand – the deployment of LAN hotspots can be accelerated by MAN, especially at locations not served by xDSL or other cable-based technologies, and the phone companies cannot provide broadband services at short notice.
- Cellular operator backhaul – these operators usually use wired and leased connections from third-party service providers and they could be replaced by MAN.
- Residential broadband – in many areas it is difficult to provide wired broadband services.
- Wireless services to rural and scarcely populated areas.
- Wireless everywhere – as the number of hotspots increases, the demand for wireless services in areas not covered by LAN increases as well.

(B) LTE

LTE is a standard for wireless data communication that is capable of overcoming some limitations of GSM/UMTS (global system for mobile communications/universal mobile telecommunications system) standards. It was developed by the 3rd Generation Partnership Project (3GPP) and its first version was established in December 2008, when 3GPP Release 8 was frozen. The main motivations for LTE are the user demands for higher data rates and quality of service, the necessity of optimizing the packet-switched system, and the demands for cost and complexity reduction.

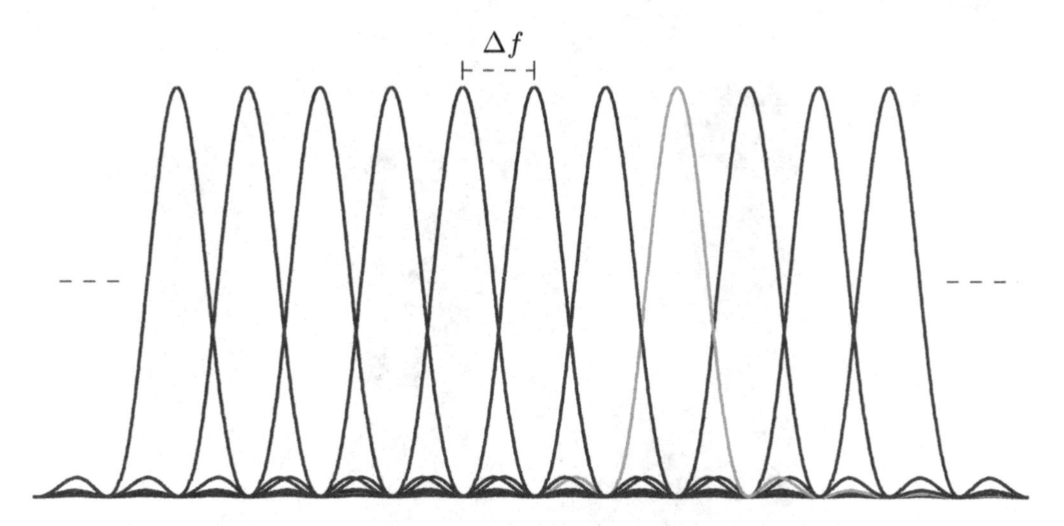

Figure 1.4: Basics of OFDM: overlapping but not interfering subcarriers.

Similar to the WiMAX, LTE uses OFDMA in the downlink connection. However, in the uplink connection, LTE opted for using the single-carrier frequency-division multiple access (SC-FDMA). In addition, LTE has very low latency and can operate on different bandwidths: 1.4, 3, 5, 10, 15, and 20 MHz.

3GPP Releases are continuously provided such as the most recent version called Release 11. It is worth mentioning that from Release 10 and beyond, the LTE is usually called LTE-Advanced, since it became IMT-Advanced compliant. Some of the key features of IMT-Advanced compliant standards are:

- Worldwide functionality and roaming;
- Compatibility of services;
- Interworking with other radio access systems;
- Enhanced peak data rates to support advanced services and applications (100 Mbps for high and 1 Gbps for low mobility).

1.3.3 BASICS OF OFDM

In the basic application of OFDM, a data stream is divided into blocks and each entry of the data block modulates a subcarrier with overlapping but not interfering frequency spectrum. As illustrated in Figure 1.4, the OFDM subcarriers are overlapped in frequency and the distance between the mainlobes of adjacent subcarriers is equal to Δf Hertz. Once a bandwidth is defined for provision of a certain service, it can be occupied by a set of OFDM subcarriers. The summation of bandlimited

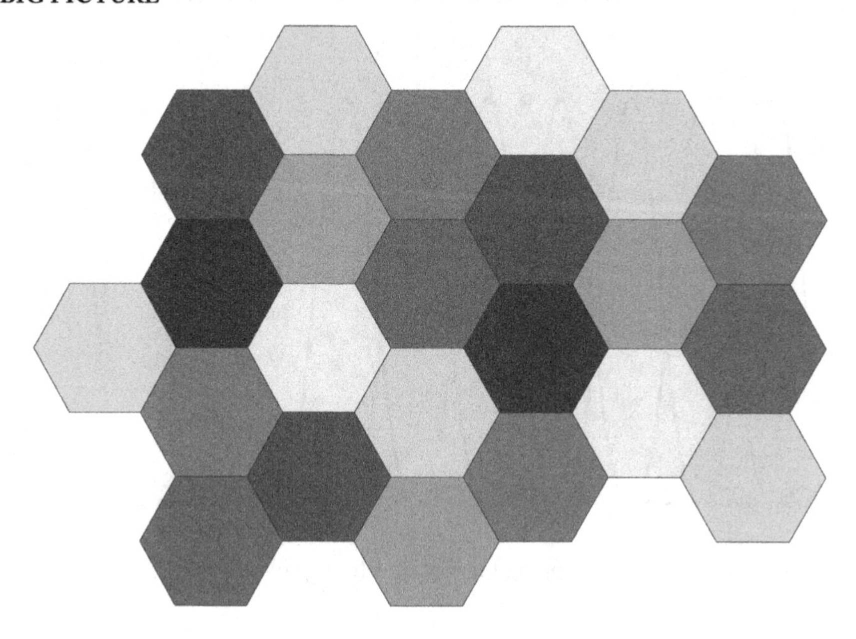

Figure 1.5: Coverage area: cellular division and frequency reuse.

OFDM subcarriers should occupy most of the available bandwidth so that the amount of symbols transmitted is maximized.

In the next section we introduce the concept of cellular division, which was crucial to the success (widespread use) of mobile communications.

1.4 CELLULAR DIVISION

The mobile communication capitalizes on the significant attenuation a transmitted signal faces in the wireless channel. This allows the wireless-service providers to reuse the available frequency range at distinct locations distant enough from each other. During wireless service deployment, the service provider defines a coverage region that consists of cell clusters. Within these clusters each cell is assigned with a different channel group.

Figure 1.5 illustrates a *coverage area* in which each *cell* is represented by a hexagon and each collection of hexagons including all colors is a *cluster*. Hexagons of the same color represent cells sharing the same radio resources, i.e., transmitting on the same frequencies. Each cluster utilizes the whole *radio resources* that can be reused by all other clusters. In actual wireless networks there are equipments, such as the mobile switching center (MSC) for GSM systems and the serving GPRS support node (SGSN) for UMTS, which dynamically distribute the radio resources among the users. These equipments can also manage an exchange of wireless channels among cells and among clusters.

When the mobile moves toward the frontier of two cells, the signal connection with the original cell becomes weak. Before it reaches the minimum level of acceptance in quality, a *handoff* to the neighboring cell should be made without interruption.

Note that the cellular division enables different users to transmit simultaneously provided they are in different cells. But, if the users are in the same cell, then a multiple access scheme must be used in order to allow simultaneous transmission for these users.

1.5 MULTIPLE ACCESS METHODS

Multiple access (MA) methods enable multiple users to transmit over the same channel. Indeed, the choice of the MA method determines how the *radio resources*, sometimes called channel resources, are shared among users. The fundamental idea underlying MA methods is the concept of *separability* among users, which means that the signals transmitted by different users, although sharing the same transmission medium, should be completely separable at the receiver. Mathematically, this implies that there must exist some domain (e.g., time, frequency, space, or code) in which the waveforms corresponding to different users are orthogonal to each other. The appropriate choice of the access method is key to achieve high data-rate transmissions, thus increasing the capacity of the networks to provide multiple services to several users.

The MA methods can be classified according to the domain in which the users are separable. The main MA methods under use are:

- *Time-division multiple access* (TDMA);
- *Frequency-division multiple access* (FDMA);
- *Code-division multiple access* (CDMA);
- *Orthogonal frequency-division multiple access* (OFDMA).

These MA methods are briefly described in the following subsections. Note, however, that they are not mutually exclusive. Indeed, the most successful 2G system, the GSM, is a hybrid of TDMA and FDMA. In GSM, each 200 kHz of channel bandwidth is shared among 8 users using TDMA.

1.5.1 TDMA

In TDMA, the separability among users occurs in the time domain, i.e., each user receives a time slot (or a set of them) for his transmission. Figure 1.6 illustrates two users sharing the same medium using a TDMA scheme. The duration of each slot is T seconds, and each user receives two of them. During the slots that *User 1* is transmitting, *User 2* is in silence, and vice versa. The signal that arrives at the receiver, assuming the channel does not introduce any distortion to the transmitted signals, is the superposition of the signals sent by the two users. Therefore, in TDMA the transmitters must be tightly synchronized in order to avoid interference among them. It is clear that the receiver must be synchronized with the transmitters as well, since it must know how to properly chop the received signal in order to isolate the signals sent by the different users.

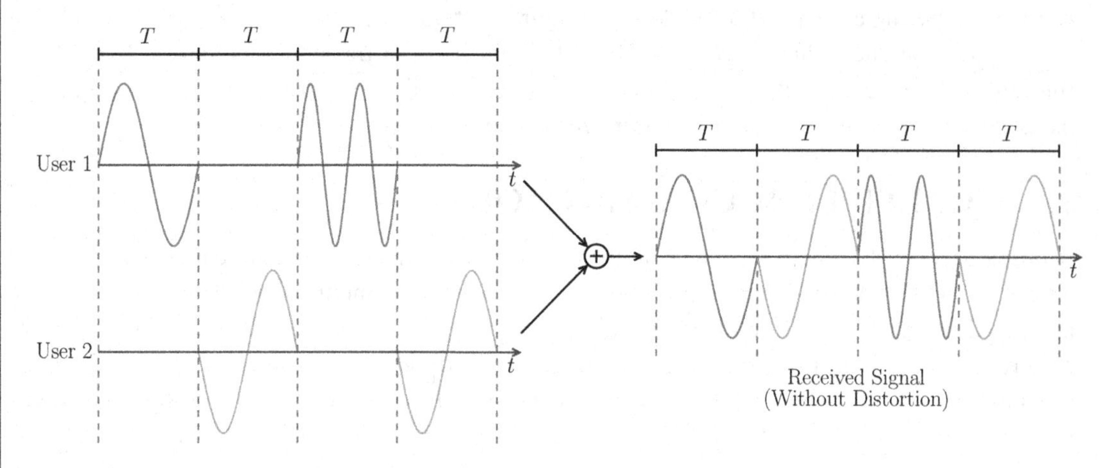

Figure 1.6: Example of TDMA involving two users.

1.5.2 FDMA

Unlike TDMA, in FDMA users may transmit all the time. The separability among them occurs in the frequency domain, i.e., for each user it is assigned a different frequency band. Figure 1.7 depicts two users employing an FDMA scheme to share the channel resources. Both users transmit over the channel using a bandwidth of Δf Hz, but the central frequency of *User 1* is f_1, whereas it is f_2 for *User 2*. Clearly, f_1 and f_2 must be distant enough from each other in order to avoid interference between the two users. At the receiver side, the signals belonging to the two users can be separated through bandpass filters that dramatically attenuate all frequencies, except the ones that fall within the desired band.

As the number and types of services available to users increase, the fixed channel assignment inherent to FDMA and TDMA becomes less efficient. As a result, more spectrum would be required with such fixed assignment. Even if some kind of flexible channel assignment is incorporated in the current FDMA and TDMA schemes, there is always a fixed upper bound in the number of users that can be served.

1.5.3 CDMA

The CDMA, sometimes called *spread-spectrum*, appears in two distinct forms: the direct sequence CDMA (DS-CDMA) and the frequency-hopping CDMA (FH-CDMA).

In DS-CDMA, the spectrum of the baseband signal is spread to occupy a wider bandwidth, as depicted in Figure 1.8. In this figure, the power spectrum density (PSD) is represented as a function of the frequency. Before transmitting the waveform conveying information, this waveform

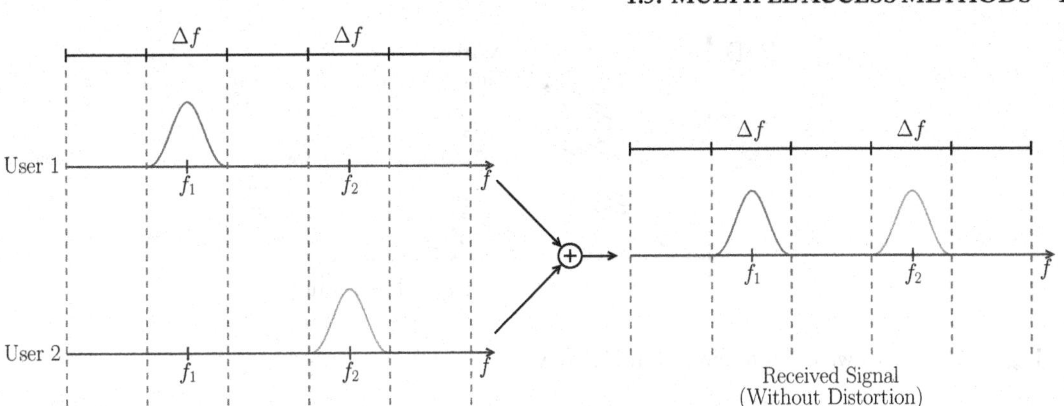

Figure 1.7: Example of FDMA involving two users.

is modified in order to spread it over a wider frequency band. At the receiver, the reverse process is performed and the spread-spectrum waveform is converted back to the original waveform from which the information can be extracted.

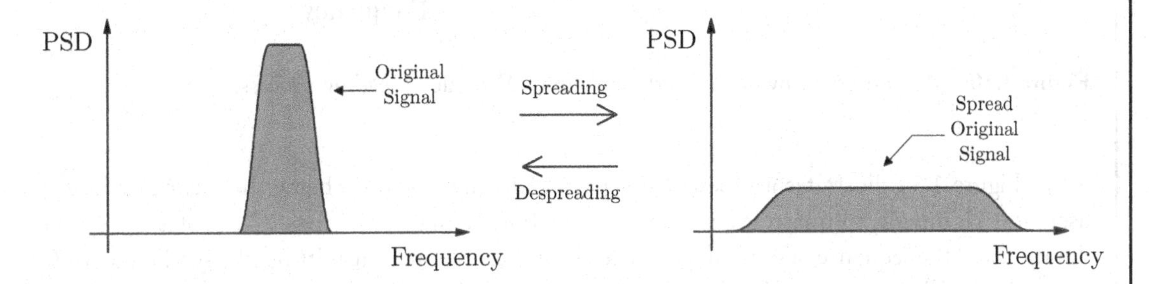

Figure 1.8: Spreading the spectrum in CDMA.

One of the main benefits from using spread-spectrum is reducing narrowband interference. For instance, consider Figure 1.9 in which the spread original signal suffers interference due to two narrowband interferers. In this case, the de-spreading process performed at the receiver de-spreads the original signal and also spreads the interferers, as depicted in Figure 1.10. Consequently, a significant portion of the signal power corresponding to the interferers is spread over frequency bands different from the ones used by the original signal, thus reducing the inband interference.

Another main advantage of CDMA is that it allows users to share all the available bandwidth simultaneously. Hence, since the signal transmitted by each user is wideband, CDMA must resolve such type of wideband interference in order to guarantee the separability among users. Indeed, CDMA introduces a new domain, called *code* domain, in which multiple users can be fully separated.

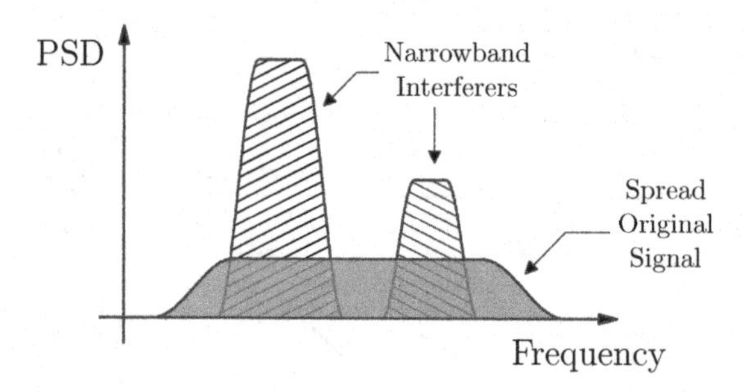

Figure 1.9: Narrowband interference in CDMA.

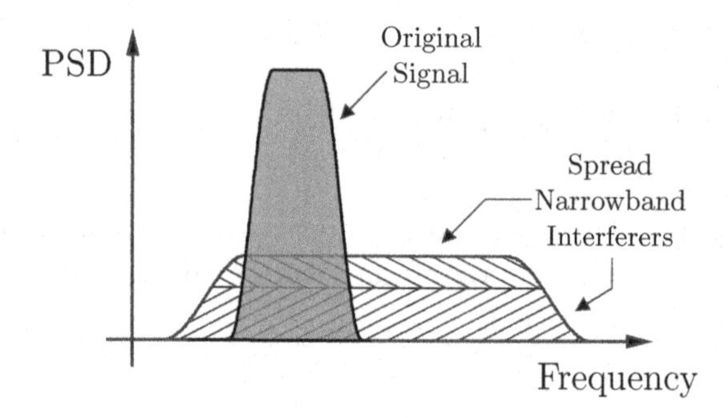

Figure 1.10: Mitigating narrowband interference in CDMA de-spreading process.

Figure 1.11 illustrates a DS-CDMA passband-transmission scheme. In this case, the mth user, $m \in \mathbb{N}$, intends to transmit a sequence of symbols $b_m(nT_b)$, in which T_b denotes the symbol duration and the sequence of symbols is indexed by $n \in \mathbb{Z}$. Before transmitting, the symbols $b_m(nT_b)$ are multiplied by a code sequence $c_m(kT_c)$, where $k \in \mathbb{Z}$ represents the sequence index and T_c corresponds to the duration of each element of the code sequence. In addition, $c_m(kT_c)$ is a very simple sequence whose elements are 1 or -1. The main difference between the sequences $b_m(nT_b)$ and $c_m(kT_c)$ is that the sampling rate of the latter is much higher, i.e., $T_c \ll T_b$. Hence, the signal resulting from the multiplication of $b_m(nT_b)$ by $c_m(kT_c)$ is a spread-spectrum signal whose bandwidth is increased by a factor equal to T_b/T_c, also known as *spreading factor* or *processing gain*. Finally, this resulting signal modulates the carrier $\cos(\omega_c t)$, where ω_c is the carrier frequency, resulting in a passband signal ready to be transmitted. The key point that guarantees the separability among users is that these code sequences are *unique*, i.e., a different code sequence is assigned to each user, and they are also *orthogonal* to each other.[4]

[4]In fact, due to some transmission issues, such as synchronization, orthogonal code sequences are sometimes exchanged by (or used together with) pseudonoise sequences.

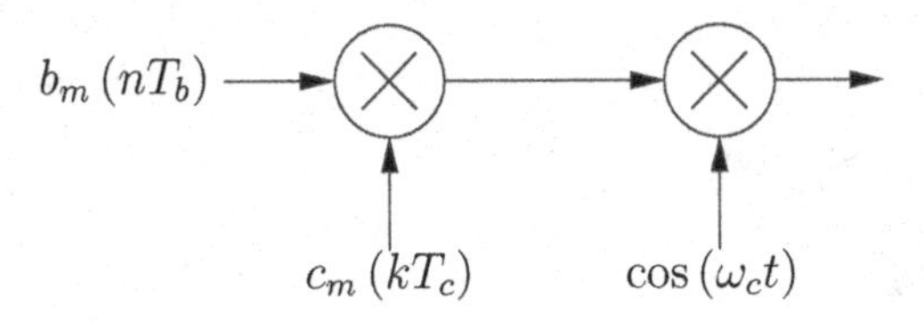

Figure 1.11: DS-CDMA passband-transmission scheme.

Note that while TDMA and FDMA have a maximum number of simultaneous users, CDMA does not have such a constraint. Nevertheless, in CDMA, if the number of users increases the service smoothly degrades.

In cellular systems, CDMA allows soft handover between neighboring cells. Close to the frontier, where the signal is weaker, the user communicates with two base stations simultaneously so that the diversity helps compensating for signal degradation at cell edges. In TDMA and FDMA neighboring cells must use different frequencies because they control interference based on spatial attenuation of the signals (frequency reuse). In CDMA, all cells use the same frequency range, eliminating the necessity for frequency-use planning, whereas TDMA and FDMA may use adaptive frequency reallocation. In addition, CDMA requires strict power control and base station synchronization, and allows intercell interference to be suppressed at the receiver.

Another type of spread-spectrum technique is the FH-CDMA, in which each user employs a different frequency band within a given time frame. During transmission, the user frequency band hops to different bands according to a prescribed hopping pattern (code). The receiver hops synchronously with the transmitter with the knowledge of the code. The hopping can be slow, where hopping occurs at the symbol rate, or fast where more than one hop occurs during symbol duration. The latter case is more difficult to implement.

As a rule, wideband MA schemes can operate in the frequency range of existing narrowband services and allow flexibility in the number of users and services provided to each user. They also allow improved interference rejection, originated from multiuser, multipath, and narrowband interferences. On the other hand, wideband MA systems require more advanced technology for implementation.

Figure 1.12 summarizes how the radio resources are assigned to one user for TDMA, FDMA, and CDMA schemes. As we have already discussed, the separability among the multiple users occurs in one of the following domains: time, frequency, or code domains. Therefore, in TDMA the user receives a time interval for his transmission, which may use the whole channel bandwidth. In FDMA, a frequency band is dedicated for the user's transmission. In CDMA, a single user may transmit all the time and using the entire channel bandwidth, but with a unique code.

1.5.4 OFDMA

In some standards, sets of OFDM subchannels can be assigned to distinct users leading to an MA scheme known as orthogonal frequency-division multiple access (OFDMA). In the simplest case,

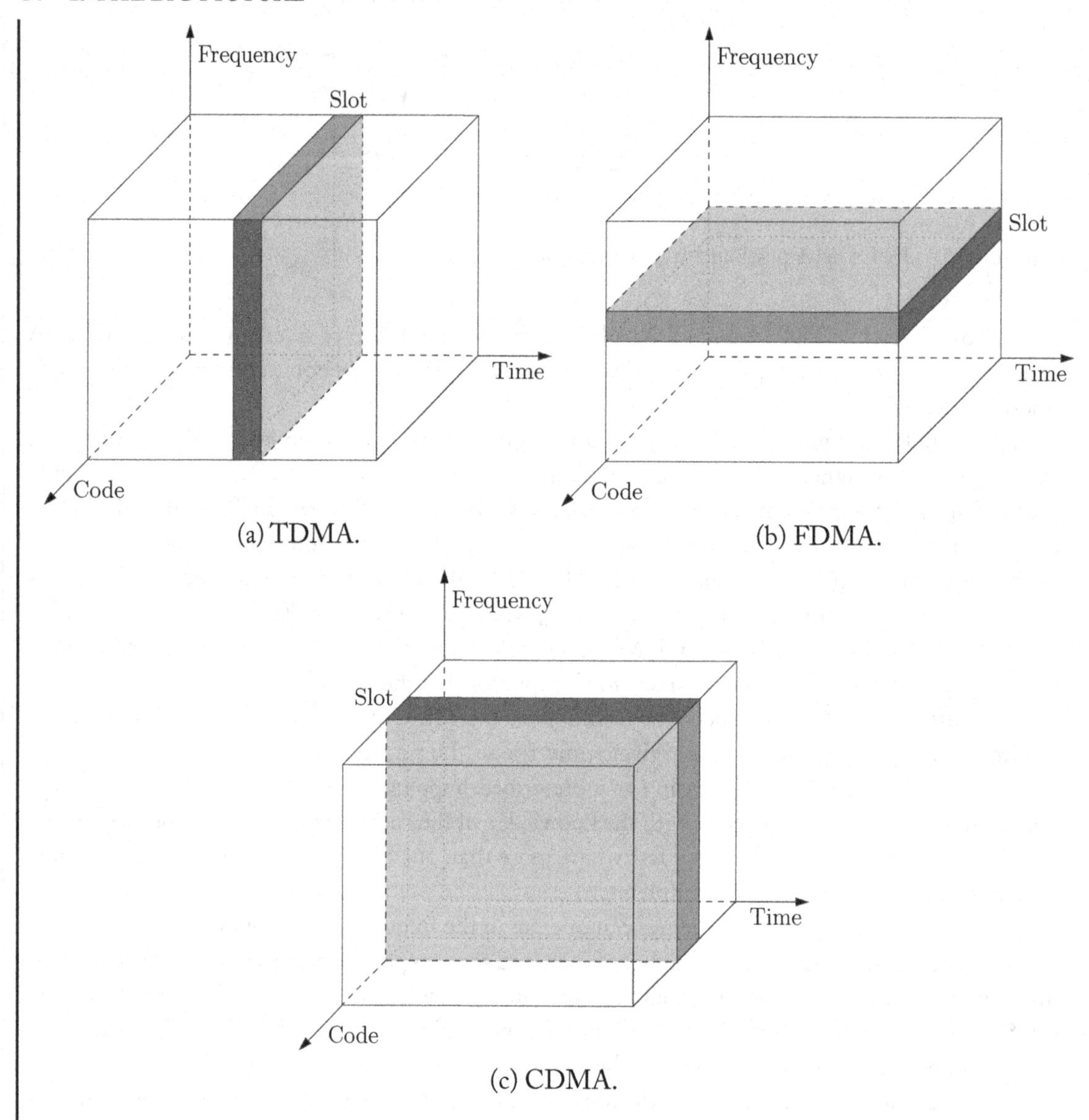

Figure 1.12: Channel sharing for: (a) TDMA; (b) FDMA; (c) CDMA.

multiple access can be implemented in a TDMA format, where at a given time slot a specific user is allowed to employ all subchannels for his transmission, as illustrated in Figure 1.13. As depicted in Figure 1.14, it is also possible to assign distinct frequency bands for different users provided the users know in which subchannels (bands) they can transmit at a given time slot.

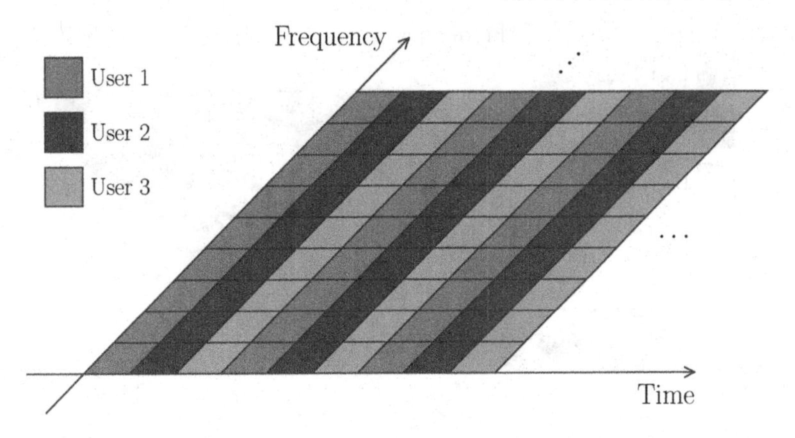

Figure 1.13: OFDMA system: TDMA case.

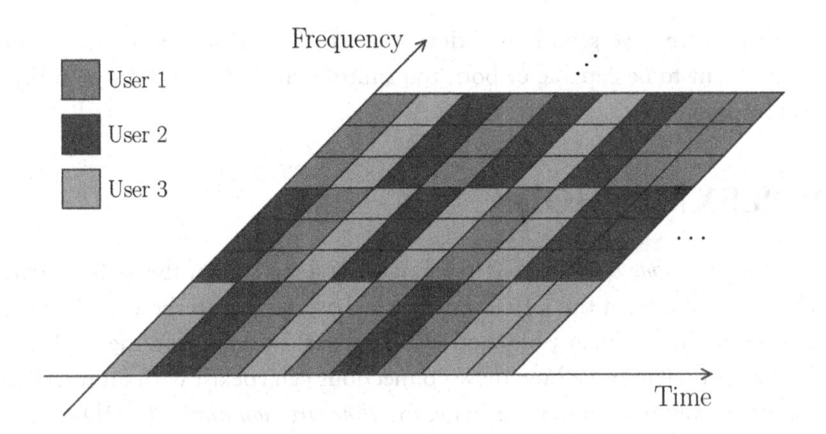

Figure 1.14: OFDMA system: FDMA case.

The most efficient way to assign the subchannels to multiple users is through random assignment, which guarantees that all users enjoy approximately the same quality of service. Figure 1.15 illustrates an OFDMA scheme with random assignment of subchannels. As depicted in this figure, the subchannels used by each user may change at each time slot and, therefore, this kind of OFDMA scheme avoids users to get stuck in low-quality subchannels. Note, however, that the OFDMA schemes that are usually employed in standards are the ones depicted in Figures 1.13 and 1.14. This is justified by some issues and limitations concerning the OFDMA with random assignment. For example, some multiple-antenna transmission schemes, especially those transmissions with diversity, cannot be employed in OFDMA with random assignment, since they usually require the transmission over adjacent subchannels.

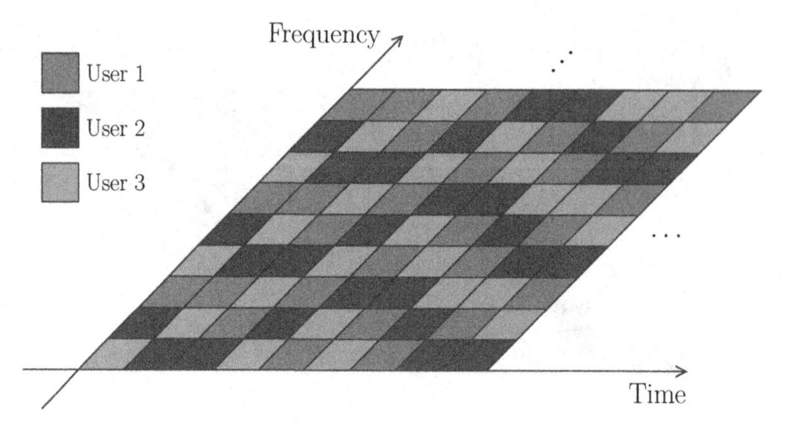

Figure 1.15: OFDMA system: random assignment.

In this section we have seen how different users can transmit over the same medium. But what if the users want to be capable of both transmitting and receiving data? In this case, duplex methods must be used.

1.6 DUPLEX METHODS

In cellular systems, *duplex methods* are used to separate the signal sent by the mobile to the base station, called *uplink* connection, from the signal sent by the base station to the mobile station, known as *downlink* connection. In fact, in any bidirectional communication system the duplex method has to be specified in order to determine how these connections can coexist without interfering with each other.[5] There are mainly two duplex methods, the *time-division duplex* (TDD) and the *frequency-division duplex* (FDD), which are briefly explained in the following.

1.6.1 TDD

As illustrated in Figure 1.16, TDD schemes assign different time intervals to uplink and downlink connections. Both of these connections can use the whole channel bandwidth during their transmissions. It is common practice to separate the uplink and the downlink connections in TDD by a time interval known as *guard time*, which avoids interference between these two connections that might be caused by propagation effects such as multipath, which are addressed in the next section.

[5]That is, bidirectional communications systems also have uplink and downlink connections like cellular systems. The difference can be on the elements that are at the ends of these connections, which can be computers and Internet service providers instead of mobile stations and base stations.

1.6.2 FDD

Figure 1.17 illustrates the FDD method in which different frequency bands are assigned to the uplink and downlink connections.

1.7 WIRELESS CHANNELS: FADING AND MODELING

The signals (waveforms) transmitted by wireless communications systems are ordinary electromagnetic waves and, as such, they suffer from wave-propagation effects. A detailed description of wireless channel effects over the transmitted signal as well as channel modeling can be found in [67, 80]. In this section, we briefly explain some of the main problems introduced by wireless channels, emphasizing the *multipath fading*, in order to motivate the mathematical model used throughout this book. These main impairments are: path loss, shadowing, and multipath fading. The problem of multiuser interference (MUI) associated with multiple access schemes is not addressed here.

1.7.1 FADING

The main effects on transmitted signals inherent to wireless channels can be summarized in one word: *fading*. Fading is a phenomenon concerning the time-variation of the channel strengths. If these variations are due to transmissions over long distances (ranging from hundreds to thousand meters), then they are known as *large-scale fading*, whereas the term *small-scale fading* is used for channel variations due to relative movements of transmitter/receiver over short distances (of the order of the carrier wavelength).

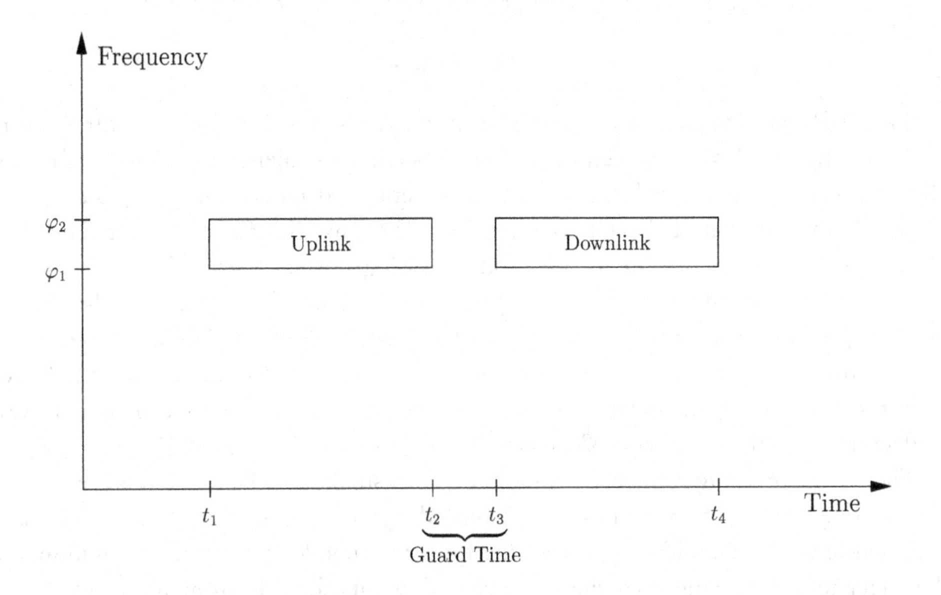

Figure 1.16: Time-division duplex (TDD).

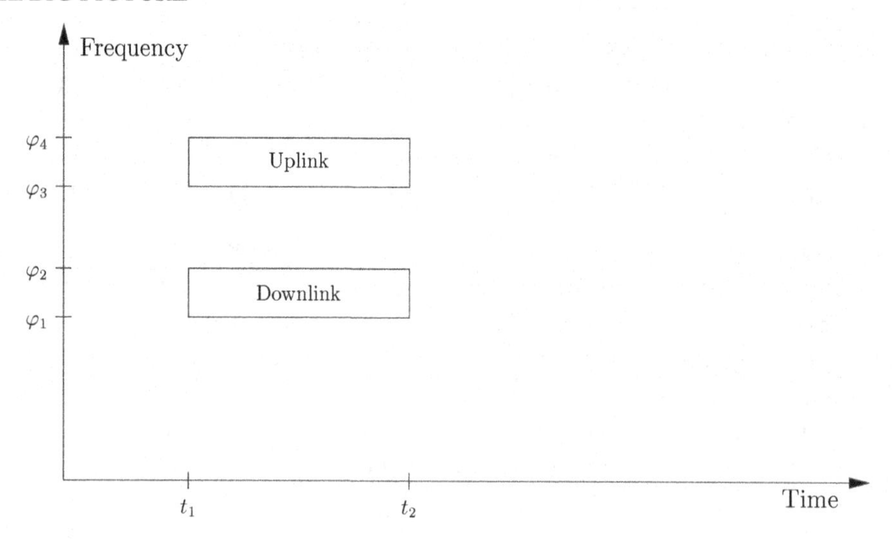

Figure 1.17: Frequency-division duplex (FDD).

Some examples of wave-propagation effects that fall into the category of large-scale fading are the *path loss* and the *shadowing* effects. *Path loss*, also known as path attenuation, is the power reduction of an electromagnetic signal as it propagates through the medium, i.e., longer distances between transmitter and receiver leads to lower power of the received signal, assuming that the power of the transmitted signal does not vary. The path loss P_{loss} is defined as

$$P_{\text{loss}} \triangleq \frac{P_t}{P_r}, \tag{1.1}$$

where P_t and P_r stand for the power of the electromagnetic signal at the transmitter and receiver, respectively. The path loss, measured in decibels, has a linear dependence on $\log_{10}(d)$, where d is the distance that the signal travels. The other effect mentioned above, namely *shadowing*, is a fading caused by the obstruction of the line-of-sight between transmitter and receiver. The shadowing generates signal variations that are usually modeled by log-normal distribution.

There are two main types of small-scale fading: the *frequency-selective* and the *flat* fading. If different frequency components of the electromagnetic signal are affected differently by the channel, such effect corresponds to *frequency-selective fading*. Otherwise, in cases the variation induced on the signal by the channel does not depend on/vary with the frequency of the electromagnetic wave, then such effect falls into the category of *flat fading*.

The *multipath fading* is the most common type of small-scale fading that is present in mobile communications. It originates from the transmitted-signal reflections in local buildings, hills, or structures around a few hundred wavelengths from the mobile. Therefore, the multipath fading can be understood as variations on the signal caused by interferences from attenuated and delayed versions of the same signal (the reflections). Figure 1.18 illustrates an example in which two reflected signals are interfering with the signal received from the line-of-sight. This figure depicts the three waveforms independently, but it is clear that the received signal $y(t)$ is a summation of the three signals. Note also that in the time interval $[0, T_s]$, in which T_s stands for the duration of one symbol generated by the digital modulator, the interference among the three signals will be constructive, whereas some destructive interferences occur within other intervals, such as $[T_s, 2T_s]$.

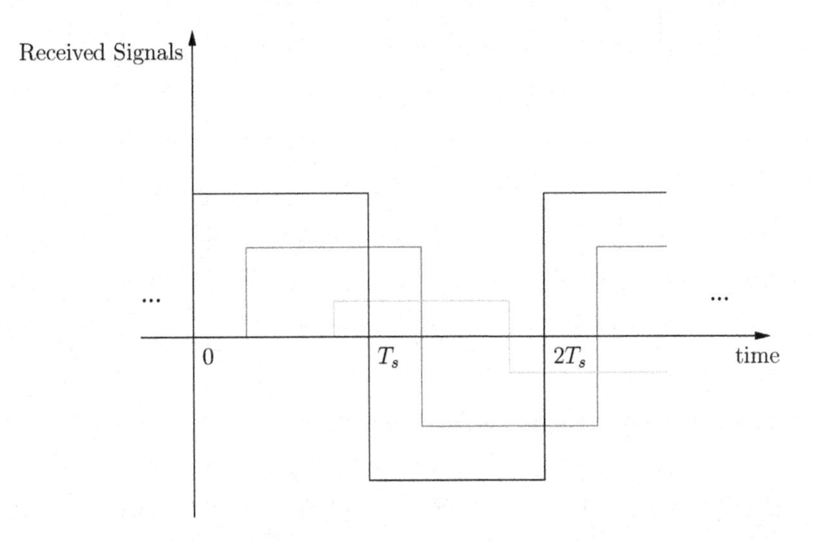

Figure 1.18: Example of multipath fading.

Figure 1.19 depicts an example of power loss in wireless environment. For a given transmitted power P_t, such figure illustrates how the received power P_r varies as a function of the transmitter-receiver distance d. These power variations occur due to the path loss, shadowing, and multipath fading phenomena.

1.7.2 MODELING

In rough environments, such as urban areas, frequency-selective fading occurs due to the reflections of the transmitted signal that arrive with distinct delays at the receiver (multipath). In these cases, the propagation medium (channel) is usually modeled as a linear system with memory, which can be characterized by its impulse responses. Indeed, since wireless channels are time-varying, their impulse responses $h(t, \tau)$ can change along the time t. Thus, the signal arriving at the receiver at a time instant $t \in \mathbb{R}$ can be written as

$$y(t) = \int_0^t s(\tau)h(t, \tau)d\tau, \tag{1.2}$$

where $h(t, \tau)$ corresponds to the channel response to an impulse applied at the instant t and $s(\tau)$ represents the transmitted signal (concatenation of waveforms representing symbols). In addition,

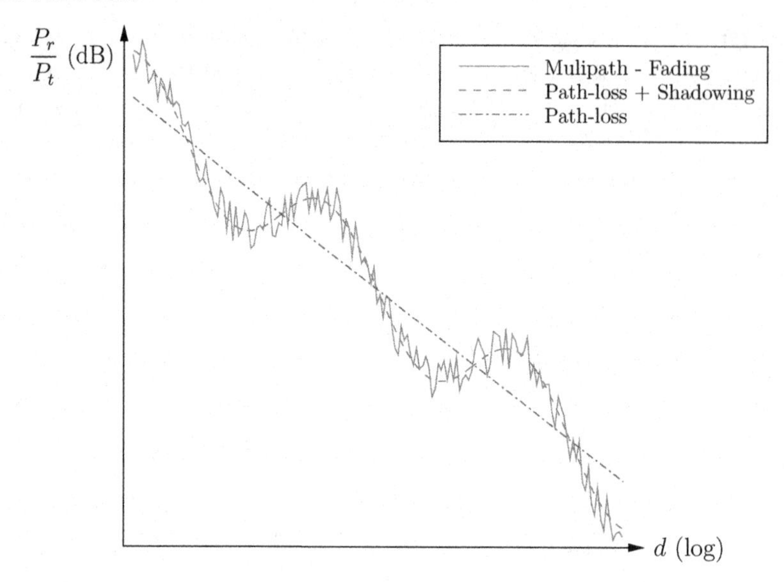

Figure 1.19: Example of power loss in wireless environment.

it is common practice to consider that the memory of $h(t, \tau)$ is finite since, due to path loss, not all existing reflections will have enough power to be sensed at the receiver.

If one is interested only in the discrete-time representation of the signals, the signal arriving at the receiver at instant $k \in \mathbb{Z}$ can be written as

$$y(k) = \sum_{i=0}^{k} s(i)h(k, i), \tag{1.3}$$

where $h(k, i)$ is the channel response to an impulse applied at instant k, and $s(i)$ represents the transmitted signal as a sequence of symbols.

Note that, in practical communications systems, the sequence of symbols $s(i)$ are usually divided into blocks, and then each block is transmitted through the channel, as will be shown in the next section.

1.8 BLOCK TRANSMISSION

Block transmission schemes are employed in some modern digital communications systems. Figure 1.20 illustrates the single-input single-output (SISO) model of a block transmission. In this figure, each data block $\mathbf{s}(n)$, where $n \in \mathbb{Z}$ represents the block index, is comprised of eight symbols coming from a digital modulator. Each data block $\mathbf{s}(n)$ is processed and prepared for transmission at the *SISO transmitter*, generating the block $\mathbf{u}(n)$ that is propagated through the *SISO channel*. If the SISO channel has memory, as usually happens in wireless channels, it will give rise to intersymbol

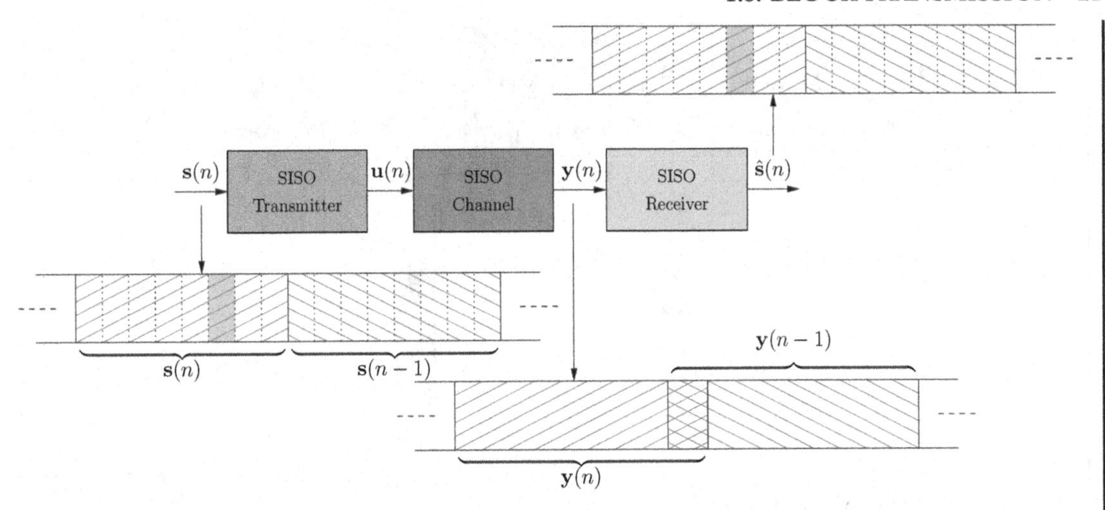

Figure 1.20: SISO model for block transmission.

interference (ISI) as well as interblock interference (IBI) at the receiver end. In this case, the currently received block $\mathbf{y}(n)$ and the previously received block $\mathbf{y}(n-1)$, corresponding to the transmitted blocks $\mathbf{u}(n)$ and $\mathbf{u}(n-1)$, respectively, will have an overlap as depicted in Figure 1.20. Then, the *SISO receiver* is responsible for eliminating both ISI and IBI and detecting the transmitted symbols, yielding a data block $\hat{\mathbf{s}}(n)$ that must be as close as possible to the originally transmitted block $\mathbf{s}(n)$.

In the case of wideband transmission, the ISI can be severe enough to make the SISO receiver very complex to implement. Roughly speaking, each transmitted symbol would spread over the time slot(s) of the neighboring symbol(s), turning their correct detection more challenging.

A naive, but widely used, solution to avoid interference among symbols of different blocks (i.e., IBI) is to allow a guard period between each block transmission. In Figure 1.20, this solution corresponds to separate each block $\mathbf{u}(n)$ by an amount of time that is sufficiently large to guarantee that overlaps do not occur between the received blocks $\mathbf{y}(n), n \in \mathbb{Z}$. The drawback of such a solution is the reduction of the data-transmission rate. This reduction can be significant in cases the blocks $\mathbf{u}(n)$ are not much larger than the guard period.

A more general block-transmission framework is illustrated in Figure 1.21. This MIMO model for block transmission encompasses many block-transmission schemes, including the SISO model, multicarrier schemes, and multiple antenna configurations. In the MIMO model for block transmission, a given data block $\mathbf{s}$ is modified by a *MIMO Transmitter*, yielding the block $\mathbf{u}$ whose length is greater than the length of $\mathbf{s}$. This larger length is due to many different reasons, such as the replication of previously transmitted blocks, in case the transmitter has memory, or just by the use of a guard period. The *MIMO channel* model can be described through a MIMO transfer function possibly with memory, generating ISI and IBI. It is up to the *MIMO receiver* to process the received signal block $\mathbf{y}$ in order to generate a reliable estimate of the transmitted signal block $\hat{\mathbf{s}}$.

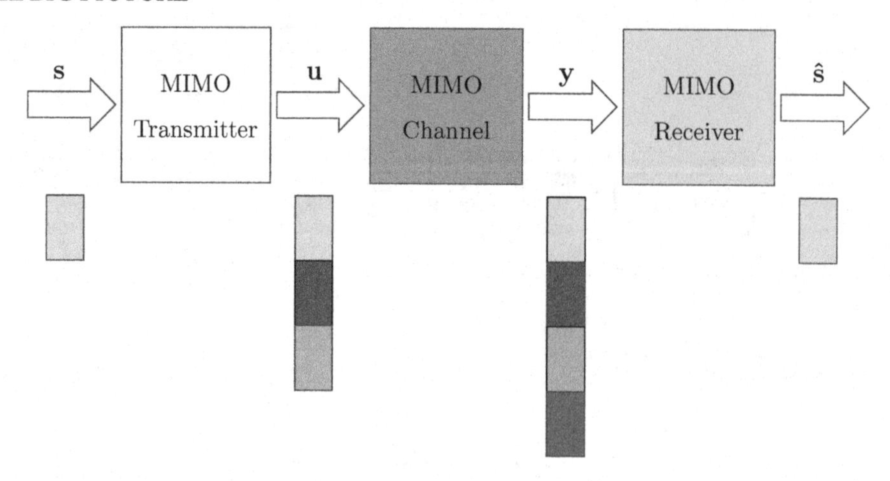

Figure 1.21: MIMO model for block transmission.

The MIMO model might represent a wide range of signal-processing tasks. In this book the main type of MIMO processing is the multicarrier transmission, which consists of transmitting each symbol in a block through a narrowband subcarrier. The benefits from using this technique are the following:

- Each subcarrier illuminates a narrow range of channel frequencies, so that the equivalent subchannel appears to be flat. This turns the equalization for each subcarrier much simpler.
- Since it consists of a block transmission, the time support for transmission of a symbol modulated by a subcarrier is roughly multiplied by the number of symbols in the block. That means there is much more time to decode the information conveyed by each subcarrier, reducing or even avoiding ISI within a block.
- If a guard period is inserted to avoid IBI, the time overhead is relatively low as long as each block carries several symbols. The guard time is a function of the length of the channel impulse response (time-delay spread).

As we have seen, if ISI and IBI are not tackled, they can deteriorate the performance of communications systems. We have also seen that IBI can be avoided just by using a simple guard period between the blocks. In the following section, we will introduce systems that tackle the ISI problem in a simple and efficient manner.

1.9 MULTICARRIER SYSTEMS

In currently deployed communications systems, multicarrier transmissions seem to be the standard choice. Multicarrier-modulation methods play a key role in modern data transmissions to deal with channels with moderate to severe intersymbol interference.

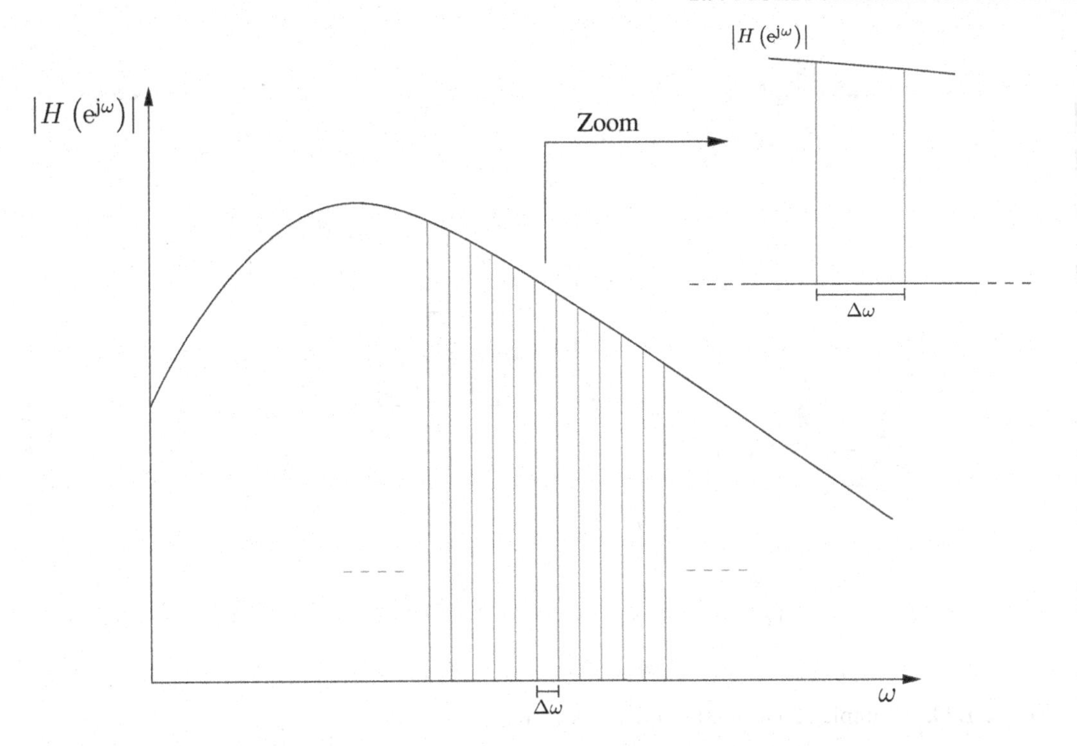

Figure 1.22: Multicarrier system dividing the channel bandwidth into non-overlapping flat subchannels.

The basic idea of multicarrier systems, whose most popular implementation is the OFDM, is the transport of information through a wideband channel by energizing it with several narrowband subcarriers simultaneously. The success of this technique relies on the partition of the physical channel into non-overlapping narrowband subchannels through a transmultiplexer, as will be explained in the next chapter. If the subchannels are narrow enough, the associated channel response in each subchannel-frequency range appears to be flat, thus avoiding the use of sophisticated equalizers. Figure 1.22 illustrates the effect of splitting a wideband channel in flatter subchannels.

Figure 1.23 depicts a transmultiplexer implementing a 4-band multicarrier system, i.e., the channel is divided into 4 subchannels. At the transmitter end, a set of symbols (represented by colors) is prepared for transmission through distinct subcarriers, where each subcarrier is represented by a finite-impulse response (FIR) filter whose transfer function is denoted by $F_i(z), i \in \{0, 1, 2, 3\}$. At the receiver side there are related FIR filters $G_i(z)$. The FIR filters can be thought as narrowband filters with distinct central frequencies, so that the symbols sent at different subcarriers travel through different subchannels, thus not interfering with each other. In addition, Figure 1.23 assumes that perfect transmission (reconstruction) is possible, i.e., the symbol transmitted at each subcarrier is perfectly recovered at the receiver.

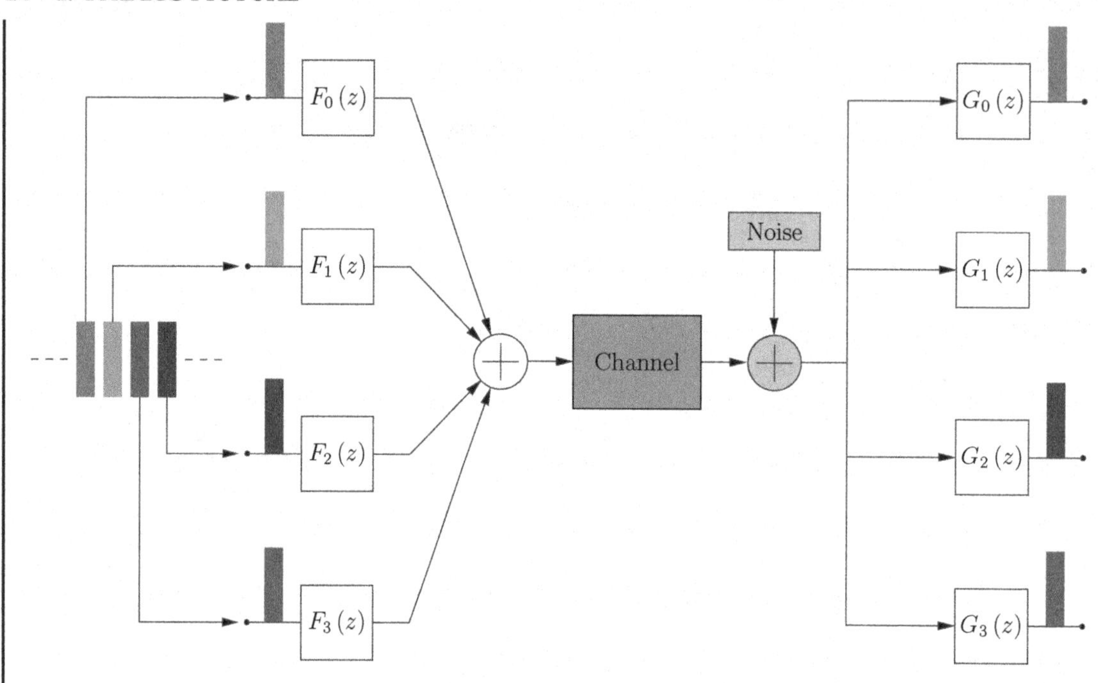

Figure 1.23: Example of a 4-band multicarrier system.

In practical systems, perfect reconstruction is usually not achievable due to degradations caused by physical-channel and noise effects, as well as power limitations. In this case, the subchannel division allows, whenever possible, the exploitation of the signal-to-noise ratio (SNR) in the distinct subbands by managing their data load in each subchannel. Indeed, if the transmitter has knowledge about the SNR at the channel output for each subcarrier, then some loading scheme could be applied, as illustrated in Figure 1.24. As can be observed in this figure, at the subcarriers with higher SNR it is possible to transmit symbols belonging to higher-order modulations, such as an 8-PSK modulation, whereas low SNR ones use lower-order modulation schemes, such as binary-PSK (BPSK). For very low SNR subcarriers, it can be even decided not to transmit any symbol at all.

In a general setup some redundancy is required at the transmission in order to keep the equalization as simple as possible. This is an important issue that will be addressed in the following chapters. In addition, several methods for jointly optimizing the transmitter and receiver of FIR MIMO systems can be employed to combat near-end crosstalk and additive-noise sources.

1.10 OFDM AS MIMO SYSTEM

In a noiseless environment, an OFDM system can be described using the MIMO framework depicted in Figure 1.21. For this case, the estimated signal vector $\hat{\mathbf{s}}$ can be described as a function of

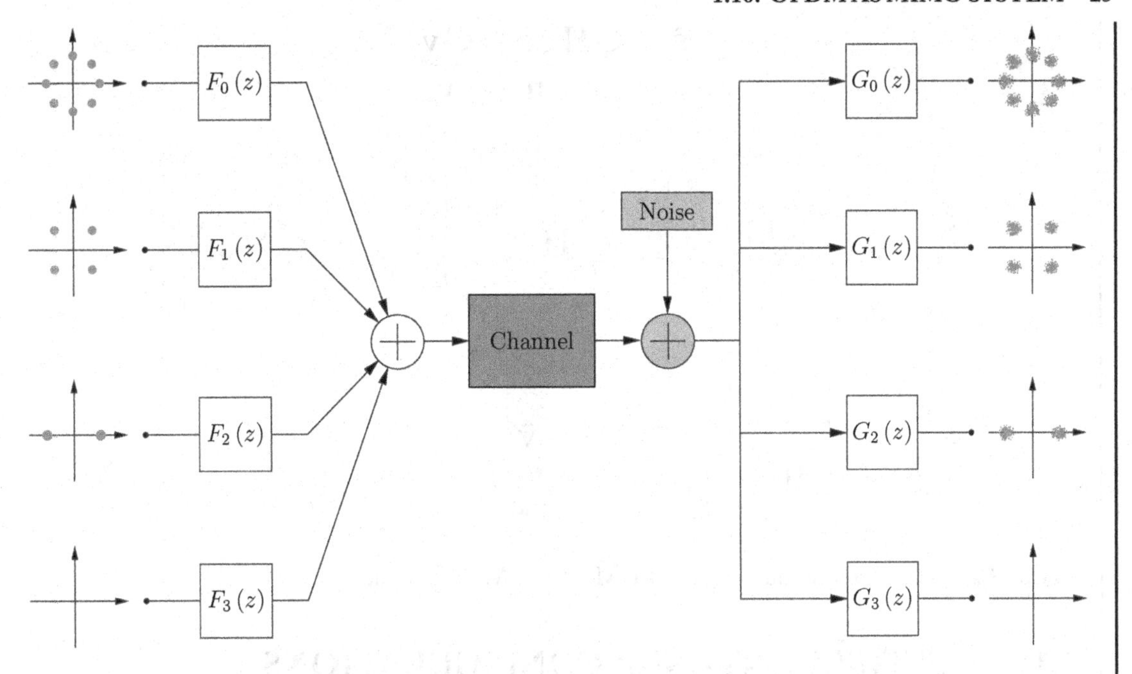

Figure 1.24: Example of a loading scheme applied to a 4-band multicarrier system.

the input-signal vector $\mathbf{s}$ as

$$\hat{\mathbf{s}} = \mathbf{GHF}\,\mathbf{s}, \tag{1.4}$$

where $\mathbf{F}$ represents the precoder matrix applied at the transmitter, $\mathbf{G}$ represents the postcoder matrix applied at the receiver, and $\mathbf{H}$ is the MIMO channel matrix. In this simplified description, all matrices are considered memoryless so that each input-signal vector is processed independently. In addition, allowing the existence of additive noise $\mathbf{v}$ at the channel output and assuming that $\mathbf{F}$ modifies and inserts a prefix on $\mathbf{s}$, the block transmission can be modeled as

$$\hat{\mathbf{s}} = \mathbf{GH}\underbrace{\mathbf{F}\,\mathbf{s}}_{\mathbf{u}} + \underbrace{\mathbf{Gv}}_{\bar{\mathbf{v}}}, \tag{1.5}$$

as represented in Figure 1.25.

For channels with memory, in which IBI exists, the OFDM system adds some redundancy to the input-signal vector in order to be able to eliminate the IBI at the receiver. As will be explained later, if the redundancy consists of a cyclic prefix and the transmitter and receiver matrices are based on discrete Fourier transform (DFT), the detection of the symbols at the receiver are decoupled from each other meaning that ISI within each block is also eliminated.

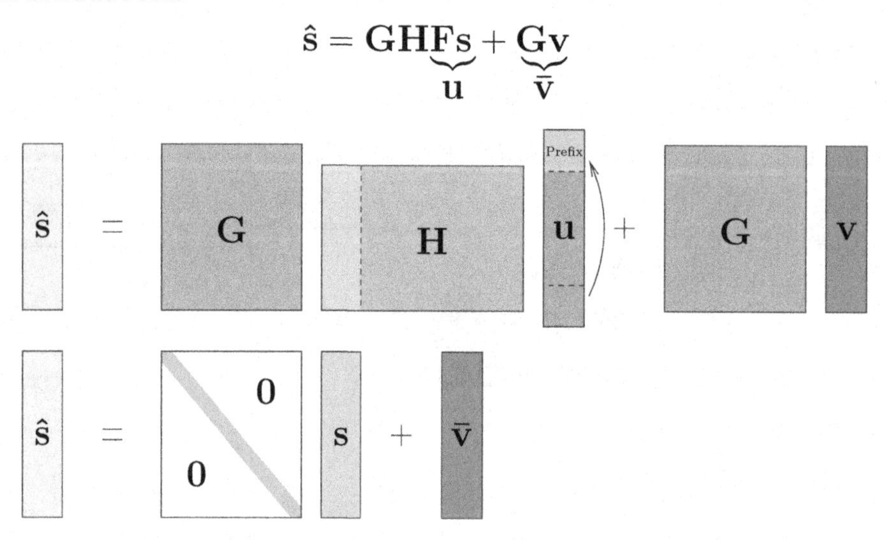

Figure 1.25: Parameter decoupling in OFDM using a MIMO model.

1.11 MULTIPLE ANTENNA CONFIGURATIONS

Although the main topic of this book is multicarrier systems, in many current applications the MIMO formulation allows the incorporation of multiple antennas at the transmitter and receiver, on top of the usual precoder and postcoder blocks inherent to these systems, as shown in Figure 1.26. These multiple-antenna building blocks introduce another degree of freedom, the *space*, that enables an efficient use of the radio resources. For instance, this new degree of freedom can be exploited to increase the system throughput by employing a *spatial multiplexing* scheme, or to enhance a transmission (i.e., decrease bit-error rate) by using a *transmission with diversity* scheme.

In a general multiple-antenna setup we can consider the transmission of several blocks of data, belonging to one or multiple users, where all the pre-processing at the transmitter is incorporated in a single matrix building block $\mathbf{T}_\text{x}$, and transmitted through an array of antennas. At the receiver there is also an array of antennas whose output signals feed a single post processing building block $\mathbf{R}_\text{x}$ that is responsible for separating and detecting each transmitted signal block, as illustrated in Figure 1.27.

There are several ways to compose the input- and output-signal vectors as well as the channel matrix in a digital communication setup. In any case, by properly stacking the transmitted and received information, the representation given by Equation (1.5) is quite powerful and accommodates several transceiver configurations.

The capacity gains of multiple-antenna systems with respect to the conventional single-antenna systems depend on:

- number of antennas at the transmitter;

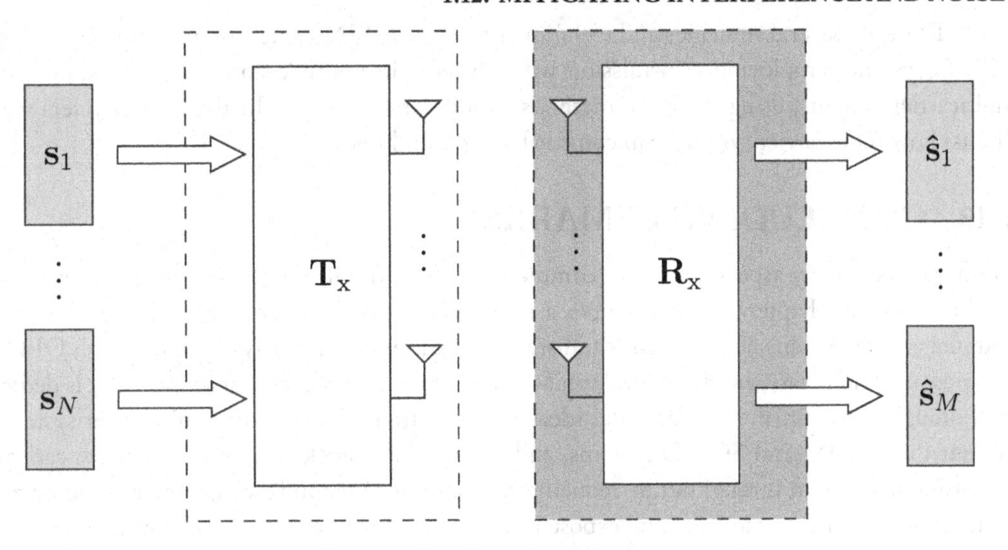

Figure 1.26: General setup of MIMO precoding with multiple antennas.

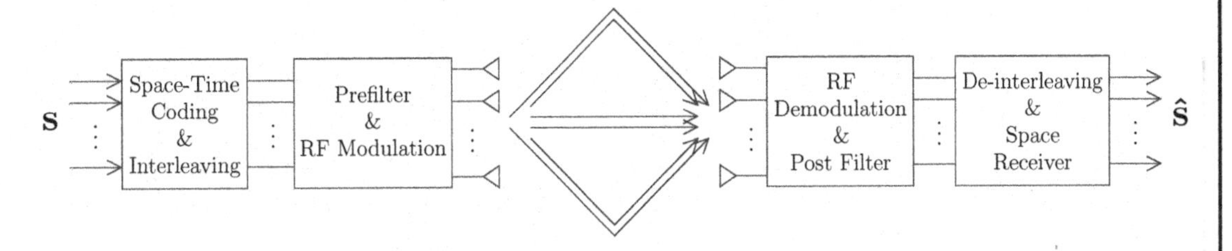

Figure 1.27: General setup of multiple block MIMO precoding with multiple antennas.

- number of antennas at the receiver;
- number of paths in the channel.

1.12 MITIGATING INTERFERENCE AND NOISE

Practical communications systems must be able to deal with interference and noise in an efficient manner. As we have already seen, mobile communications suffer ISI and IBI due to multipath fading, and they also suffer interference caused by other users sharing the same radio resources, which is usually called multiuser interference (MUI) or co-channel interference.

In this context, the signal-to-interference-plus-noise ratio (SINR) plays a key role in assessing the quality of a transmission. A high SINR, indicating a high-quality transmission, can be achieved by mitigating interference and/or noise, promoting enhancements in the performance of the physical layer.

There are several strategies and techniques to increase SINR, such as: designing equalizers for MIMO systems, employing transmission with diversity in multiple-antenna systems, optimizing multicarrier systems, using subspace methods to mitigate noise, etc. In the next chapters we will discuss how multicarrier systems can combat ISI, IBI, and noise.

1.13 CONCLUDING REMARKS

In this chapter, many aspects of digital communications and transmissions were briefly introduced. In the following chapters, some of these aspects will be used or carefully revisited. Among the material covered in this chapter, block transmissions and multicarrier systems, in which OFDM is the most notorious case, are the central topics of this book. Indeed, the rest of this book is dedicated to thoroughly explain the fundamental ideas of block transmissions and multicarrier systems, to present the OFDM and SC-FD systems, and to introduce block transceivers that are capable of increasing the system throughput by reducing the amount of redundancy necessary to remove IBI. In addition, we tried to motivate and expose in an intuitive manner the importance and necessity of such topics to current communications systems. Therefore, the approach followed here consists of presenting the material in a pictorial way, leaving the mathematical details to the following chapters.

CHAPTER 2

Transmultiplexers

2.1 INTRODUCTION

The proposal of new techniques for channel and source coding, along with the development of integrated circuits and the use of digital signal processing (DSP) for communications have allowed the deployment of several communications systems to meet the demands for transmissions with high data-rates. Typical DSP tools, such as digital filtering, are key to retrieving, at the receiver end, reliable estimates of signals associated with one or several users who share the same physical channel.

There are various classes of digital filters. Those employed in communications systems can be either fixed or adaptive, linear or nonlinear, with finite impulse response (FIR) or with infinite impulse response (IIR), just to mention a few. Among such classes of systems, fixed, linear, and FIR filters are rather common in practice because of their simpler implementation, good stability properties, and lower costs as compared to other alternatives.

Nonetheless, modern communications systems require more sophisticated techniques, thus calling for more features than fixed, linear, and FIR filters can offer. In this context, multirate signal processing adds some degrees of freedom to the standard linear time-invariant (LTI) signal processing through the inclusion of *decimators* and *interpolators*. These degrees of freedom are crucial to develop some interesting representations of communications systems based on filter banks, especially multicarrier transceivers. A *filter bank* is a set of filters (usually LTI FIR filters) sharing the same input-output pair and internally employing decimators and interpolators.

Filter-bank representations are widely used in source coding and spectral analysis. In communications, the *transmultiplexer* (TMUX) configuration can be employed to represent multicarrier or single-carrier transceivers, and can be considered a system dual to the filter-bank configuration in the sense that the signal processing which takes place at the input of a filter bank actually appears at the output of a TMUX, and *vice versa*. Indeed, several practical systems can be modeled using TMUXes.

Unlike filter banks that usually require sharp frequency-selective subfilters, practical multicarrier transceivers can be modeled as TMUXes which use short-length subfilters with poor frequency selectivity. In the majority of practical cases, these transceivers are implemented as *memoryless block-based transceivers*. The most commonly used memoryless block-based transceivers are the orthogonal frequency-division multiplexing (OFDM) and the single-carrier with frequency-domain equalization (SC-FD) systems.

The main feature of OFDM-based transceivers is the elimination of *intersymbol interference* (ISI) with low computational complexity, i.e., using just a small amount of numerical operations to undo the harmful effects induced by frequency-selective channels. A competing alternative to OFDM is the SC-FD transceiver, which presents lower peak-to-average power ratio (PAPR) and lower sensitivity to carrier-frequency offset (CFO), as explained in [63, 87]. In addition, for frequency-selective channels, the bit-error rate (BER) of SC-FD can be lower than for its OFDM counterpart, particularly for the cases in which the channel has high attenuation at some subchannels.

In this chapter some key multirate signal-processing tools are revised (Section 2.2) aiming at their use in the modeling of communications systems (Section 2.3). These tools will be particularly utilized to represent OFDM and SC-FD systems, as well as to introduce some initial results related to what is beyond OFDM-based systems, namely: the memoryless LTI block-based transceivers using reduced redundancy (Section 2.4).

2.2 MULTIRATE SIGNAL PROCESSING

In many signal-processing applications, it is quite common that signals with distinct sampling rates coexist. In general, multirate signal-processing systems include as building blocks both the interpolator and the decimator. The *interpolation* consists of increasing the sampling rate of a given signal, whereas the *decimation* entails a sampling-rate reduction of its input signal. The loss of data inherent to decimation may give rise to aliasing in the decimated signal spectrum.

The *interpolation* by a factor $N \in \mathbb{N}$ consists of including $N-1$ zeros between each pair of adjacent samples, generating a signal whose sampling rate is N times larger than the sampling rate of the original signal. Indeed, given a complex-valued signal $s(n)$, in which the integer number n denotes the time index at the original sampling rate, the interpolated signal $s_{\text{int}}(k)$ is given by

$$s_{\text{int}}(k) \triangleq \begin{cases} s(n), & \text{whenever } k = nN \\ 0, & \text{otherwise} \end{cases}, \tag{2.1}$$

where the integer number k denotes the time index at the new sampling rate. In the frequency domain, the effect of interpolation can be described as (see, for example, [17])

$$S_{\text{int}}(\mathrm{e}^{\mathrm{j}\omega}) = S(\mathrm{e}^{\mathrm{j}\omega N}), \tag{2.2}$$

in which

$$\begin{aligned} X(\mathrm{e}^{\mathrm{j}\omega}) &\triangleq \mathcal{F}\{x(n)\} \\ &= \sum_{n\in\mathbb{Z}} x(n)\mathrm{e}^{-\mathrm{j}\omega n} \end{aligned} \tag{2.3}$$

is the discrete-time Fourier transform of the sequence $x(n)$, with $\omega \in \mathbb{R}$ denoting the frequency variable.[1]

[1] It is assumed that the discrete-time Fourier transform of the sequence $x(n)$ exists, i.e., the series in expression (2.3) is convergent for all real-valued scalar ω. For instance, an absolutely summable (i.e., an ℓ_1-signal) $x(n)$ is sufficient to guarantee the convergence of the series.

The *decimation* by a factor N consists of discarding $N-1$ samples from each non-overlapping block containing N samples of the input signal. The resulting signal has a sampling rate N times lower than the sampling rate of the original signal. Indeed, given the signal $s(n)$, the decimated signal $s_{\text{dec}}(k)$ is defined by

$$s_{\text{dec}}(k) \triangleq s(kN), \tag{2.4}$$

for all integer number k. In the frequency domain, it is possible to show that the decimated signal is represented by (see, for example, [17])

$$S_{\text{dec}}(e^{j\omega}) = \frac{1}{N}\sum_{n\in\mathcal{N}} S\left(e^{j\frac{\omega-2\pi n}{N}}\right), \tag{2.5}$$

where $\mathcal{N} \triangleq \{0, 1, \cdots, N-1\}$. Unlike the interpolation, the decimation is a periodically time-varying operation.

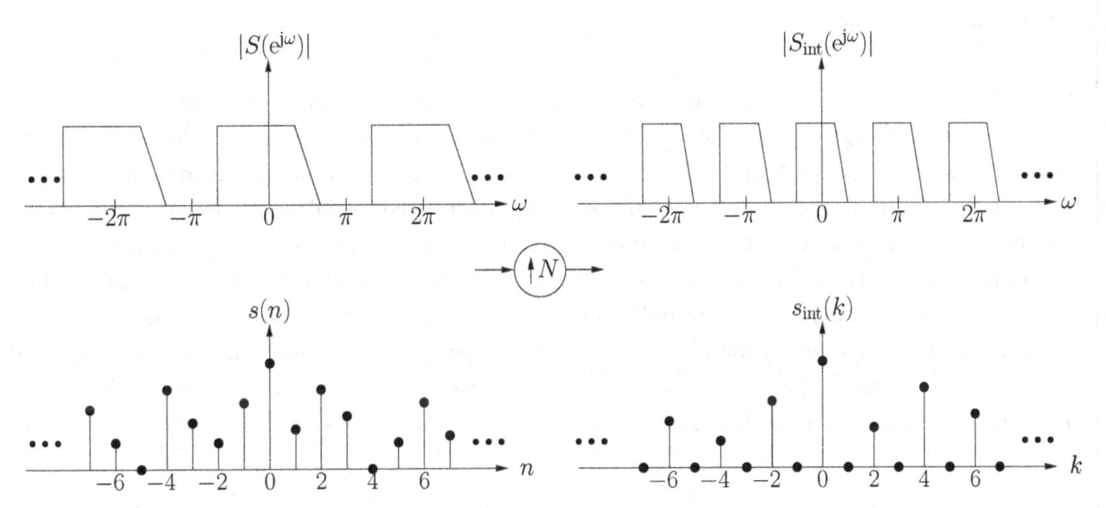

Figure 2.1: Interpolation ($N = 2$).

It is worth mentioning that a more appropriate nomenclature for the interpolation and decimation processes just described should be *upsampling* and *downsampling*, reserving the nouns interpolation and decimation for the cases in which a filtering process is also present. However, it is rather common in the literature and in practice to use interchangeably the nomenclatures upsampling/interpolation and downsampling/decimation. We will follow this practice, but the reader will be able to identify easily when a filtering process takes place or not.

Figures 2.1 and 2.2 depict the respective effects of interpolation and decimation by a factor $N = 2$ in both time and frequency domains. These signals are only for illustration purposes and they do not represent true time-frequency pairs. By examining Figures 2.1 and 2.2 it is possible to

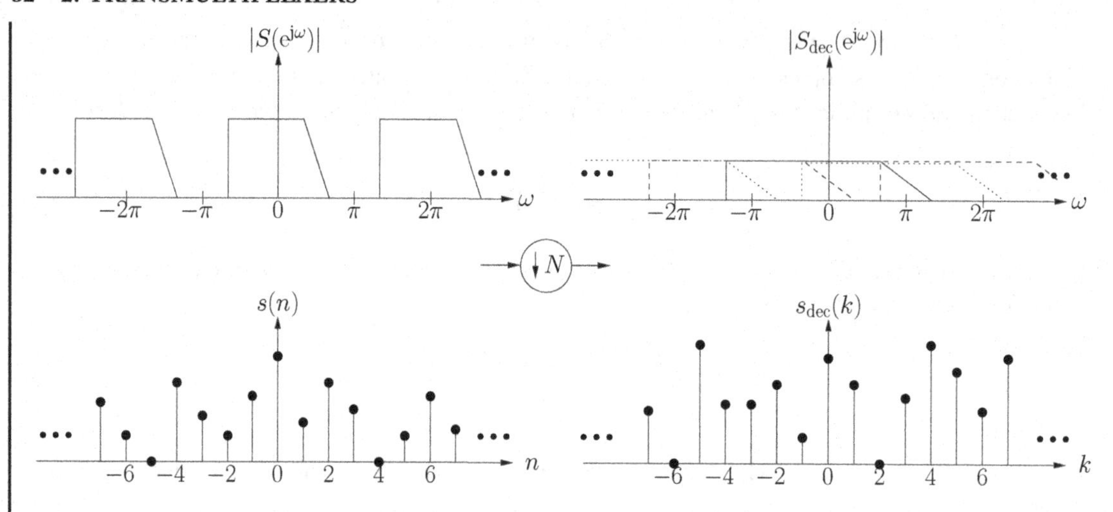

Figure 2.2: Decimation ($N = 2$).

verify that, in order to avoid aliasing due to decimation and to eliminate the spectrum repetition due to interpolation, a digital filtering operation is required before the decimation and after the interpolation. The decimation filter narrows the spectrum of the input signal in order to avoid that aliasing corrupts the spectrum of the resulting decimated signal. For a lowpass real signal, for instance, we have to maintain the input signal information only at the low frequencies within the range $\left(-\frac{\pi}{N}, \frac{\pi}{N}\right)$, so that the spectrum at this range is not corrupted after decimation. The interpolation filter smooths the interpolated signal $s_{\text{int}}(k)$, eliminating abrupt transitions between non-zero and zero samples, which is the source of the spectrum repetitions (also known as *spectral images*). The central frequencies of the spectrum repetitions are located at $\pm\frac{2\pi}{N}n$, with $n \in \mathcal{N}$. Figure 2.3 illustrates how the decimation and interpolation operations are implemented in practice.

$s(n) \rightarrow \boxed{\uparrow N} \rightarrow \boxed{f(k)} \rightarrow s_{\text{int}}(k)$ $\qquad$ $s(n) \rightarrow \boxed{g(n)} \rightarrow \boxed{\downarrow N} \rightarrow s_{\text{dec}}(k)$

Figure 2.3: General interpolation and decimation operations in time domain.

Example 2.1 (Decimation & Interpolation) Let $h(n)$ be a signal defined as

$$h(n) \triangleq \begin{cases} 2^n, & \text{whenever } n \in \{0, 1, \cdots, 7\} \\ 0, & \text{otherwise} \end{cases} . \tag{2.6}$$

Determine $H_{\text{int}}(z) = [H(z)]_{\uparrow 3}$ and $H_{\text{dec}}(z) = [H(z)]_{\downarrow 3}$, for all non-zero complex number z, in which $H(z) \triangleq \mathcal{Z}\{x(n)\}$ is the $\mathcal{Z}$-transform of the sequence $x(n)$. In addition, the notations $[(\cdot)]_{\uparrow N}$ and $[(\cdot)]_{\downarrow N}$ denote the interpolation and decimation by N applied to $(\cdot)$, respectively.

Solution. We know that

$$
\begin{aligned}
H(z) &= \sum_{n\in\mathbb{Z}} x(n)z^{-n} \\
&= 1 + 2z^{-1} + 4z^{-2} + 8z^{-3} + 16z^{-4} + 32z^{-5} + 64z^{-6} + 128z^{-7},
\end{aligned} \tag{2.7}
$$

for all $z \neq 0$. The interpolation by a factor of 3 is equivalent to insert 2 zero-valued samples between adjacent samples of $x(n)$. Hence, we have

$$
\begin{aligned}
H_{\text{int}}(z) =& 1 + 0.z^{-1} + 0.z^{-2} + 2z^{-3} + 0.z^{-4} + 0.z^{-5} + 4z^{-6} + 0.z^{-6} + 0.z^{-8} + 8z^{-9} \\
&+ 0.z^{-10} + 0.z^{-11} + 16z^{-12} + 0.z^{-13} + 0.z^{-14} + 32z^{-15} + 0.z^{-16} + 0.z^{-17} \\
&+ 64z^{-18} + 0.z^{-19} + 0.z^{-20} + 128z^{-21} \\
=& 1 + 2z^{-3} + 4z^{-6} + 8z^{-9} + 16z^{-12} + 32z^{-15} + 64z^{-18} + 128z^{-21} \\
=& H(z^3),
\end{aligned} \tag{2.8}
$$

for all $z \neq 0$. The decimation by a factor of 3 will generate a discrete-time signal $h_{\text{dec}}(k) = h(3k) = 2^{3k} = 8^k$, if $k \in \{0, 1, 2\}$, or $h_{\text{dec}}(k) = 0$, otherwise. Hence, we have

$$
H_{\text{dec}}(z) = 1 + 8z^{-1} + 64z^{-2}, \tag{2.9}
$$

for all $z \neq 0$. □

In multirate systems, there are very useful manners to manipulate the interpolation and decimation building blocks. We are particularly interested in ways to commute the decimation and interpolation operations with linear time-invariant filters. Some forms of commuting are based on the so-called *noble identities*.

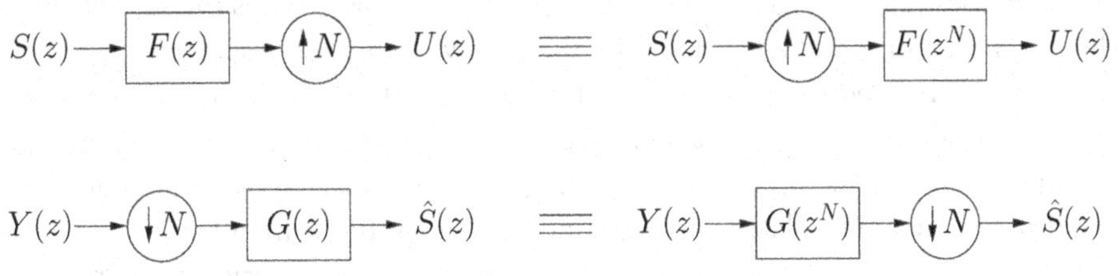

Figure 2.4: Noble identities in $\mathcal{Z}$-domain.

Figure 2.4 illustrates the building-block representations of the noble identities. In the interpolation process, instead of first upsampling the input signal and then filtering it, one can first filter the input signal in a lower sampling rate and then upsample the resulting signal. This strategy allows one to reduce the number of operations required by the overall interpolation process. As for the decimation process, the decimator followed by a filter is equivalent to filter the input signal by the interpolated filter followed by the downsampling. These operations can be mathematically described as

$$[F(z)S(z)]_{\uparrow N} \triangleq U(z) = F(z^N)\,[S(z)]_{\uparrow N}\,, \tag{2.10}$$

$$G(z)\,[Y(z)]_{\downarrow N} \triangleq \hat{S}(z) = \Big[G(z^N)Y(z)\Big]_{\downarrow N}\,. \tag{2.11}$$

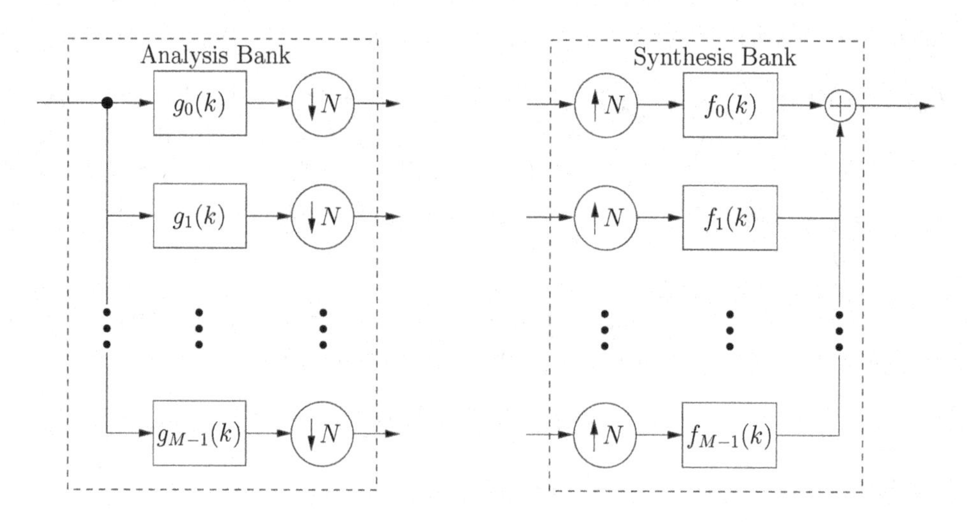

Figure 2.5: Analysis and synthesis filter banks in time domain.

A widespread application of multirate systems is the filter-bank design. A *filter bank* consists of a set of filters with the same input signal, or a set of filters whose outputs are added to form the overall output signal, as depicted in Figure 2.5. The set of $M \in \mathbb{N}$ filters represented by the family of impulse responses $\{g_m(k)\}_{m\in\mathcal{M}}$, in which $\mathcal{M} \triangleq \{0, 1, \cdots, M-1\}$, is the so-called *analysis filter bank*, whereas the set of filters represented by the family of impulse responses $\{f_m(k)\}_{m\in\mathcal{M}}$ is the *synthesis filter bank*. It is possible to verify that the analysis filter bank divides the input signal in subbands, generating narrowband signals which can be further decimated. The subband signals can be employed for analyses and manipulations according to the particular application. For reconstruction, the subband signals are interpolated and combined by the synthesis filter bank.

Transmultiplexers, also known as *filter-bank transceivers*, are considered systems dual to the filter-bank configurations since the roles of analysis and synthesis filter banks are interchanged in transmultiplexers. Indeed, the inputs of a transmultiplexer are first combined by the synthesis bank and, after some further processing stages, the outputs are obtained as a result from the analysis bank, as shown in Figure 2.6.

It is worth mentioning that this section is based on [17, 81], which contain a thorough treatment of this subject.

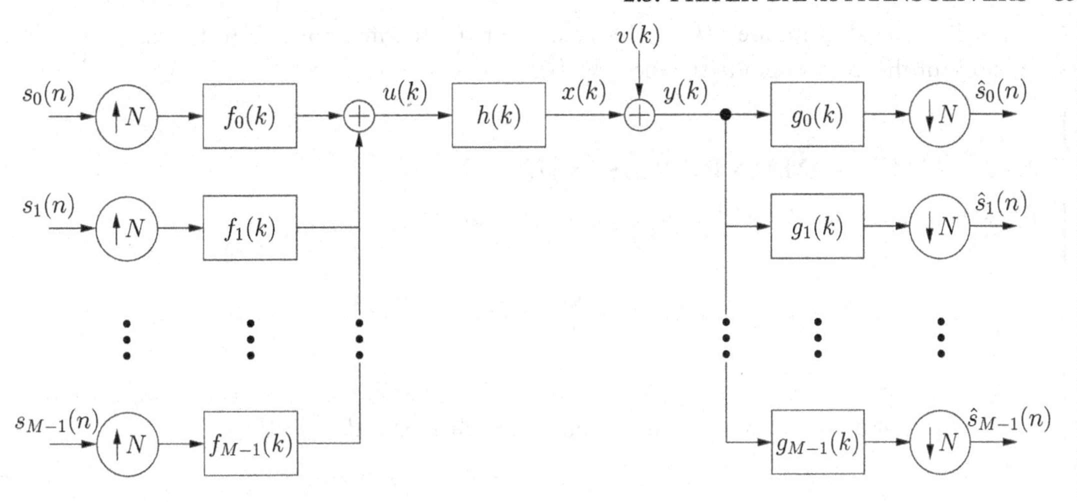

Figure 2.6: TMUX system in time domain.

2.3 FILTER-BANK TRANSCEIVERS

Consider the transceiver model described in Figure 2.6, where a communication system is modeled as a multiple-input multiple output (MIMO) system. The data samples of each sequence $s_m(n)$ belong to a particular digital constellation $\mathcal{C} \subset \mathbb{C}$, such as PAM, QAM, or PSK.[2] The sequence $s_m(n)$ represents the mth transceiver input, where $m \in \mathcal{M}$ and $n \in \mathbb{Z}$ represents the time index. The corresponding transceiver output is denoted as $\hat{s}_m(n) \in \mathbb{C}$, which should be a reliable estimate of $s_m(n-\delta)$, where $\delta \in \mathbb{N}$ represents the delay introduced by the overall transmission/reception process.

A communication system can be properly designed by carefully choosing the set of causal transmitter filters with impulse responses represented by $\{f_m(k)\}_{m\in\mathcal{M}}$, and the set of causal receiver filters represented by $\{g_m(k)\}_{m\in\mathcal{M}}$. These filters operate at a sampling rate N times larger than the sampling rate of the sequences $s_m(n)$. Note that the index n represents the sample index at the input and output of the transceiver, whereas $k \in \mathbb{Z}$ is employed to represent the sample index of the subfilters and of the internal signals between the interpolators and decimators. The transmitter and receiver subfilters are time invariant in our discussions.

The input signals $s_m(n)$, for each $m \in \mathcal{M}$, are processed by the transmitter and receiver subfilters aiming at reducing the channel distortion, so that the output signals $\hat{s}_m(n)$ may represent good estimates of the corresponding transmitted signals. The usual objective in a communication system is to produce estimates of $s_m(n-\delta)$ achieving low BER and/or maximizing the data throughput.

The channel model can be represented by an FIR filter of order $L \in \mathbb{N}$ whose impulse response is $h(k) \in \mathbb{C}$. The FIR transfer function accounts for the frequency-selective behavior of the physical

[2]Pulse-amplitude modulation, quadrature-amplitude modulation, or phase-shift keying, respectively.

channel. The additive noise $v(k) \in \mathbb{C}$ accounts for the thermal noise from the environment and possibly for the multi-user interference (MUI).

2.3.1 TIME-DOMAIN REPRESENTATION

According to Figure 2.6 the channel input signal is given by

$$u(k) \triangleq \sum_{(i,m)\in\mathbb{Z}\times\mathcal{M}} s_m(i) f_m(k - iN). \tag{2.12}$$

The relation between input and output of the channel is described as

$$y(k) \triangleq \sum_{j\in\mathbb{Z}} h(j)u(k - j) + v(k). \tag{2.13}$$

At the receiver end, the signal $y(k)$ is processed in order to generate estimates of the transmitted data according to

$$\hat{s}_m(n) \triangleq \sum_{l\in\mathbb{Z}} g_m(l)y(nN - l). \tag{2.14}$$

By employing Equations (2.12), (2.13), and (2.14) it is possible to describe the relation between the input signal $s_m(n)$ and its estimate $\hat{s}_m(n)$, as given by

$$\hat{s}_m(n) = \sum_{(i,j,l,m)\in\mathbb{Z}^3\times\mathcal{M}} g_m(l)h(j)s_m(i) f_m(nN - l - j - iN) + \sum_{l\in\mathbb{Z}} g_m(l)v(nN - l). \tag{2.15}$$

The description above is not the easiest one to analyze the system and draw conclusions. For example, it is possible to employ a time-domain approach using matrix description, as described in [70, 72]. Another approach is to apply polyphase decomposition in a $\mathcal{Z}$-domain formulation as described as follows.

2.3.2 POLYPHASE REPRESENTATION

As long as the interpolation and decimation factors are equal to N, it is convenient to describe the transmitter and receiver filters by their *polyphase decompositions* of order N, according to the expressions

$$
\begin{aligned}
F_m(z) &\triangleq \mathcal{Z}\{f_m(k)\} \\
&= \sum_{k\in\mathbb{Z}} f_m(k) z^{-k} \\
&= \sum_{j\in\mathbb{Z}} \left[f_m(jN)z^{-jN} + f_m(jN+1)z^{-(jN+1)} + \cdots + f_m(jN+N-1)z^{-(jN+N-1)} \right] \\
&= \sum_{j\in\mathbb{Z}} \sum_{i\in\mathcal{N}} f_m(jN+i) z^{-(jN+i)} \\
&= \sum_{i\in\mathcal{N}} \sum_{j\in\mathbb{Z}} f_m(jN+i) z^{-jN} z^{-i} \\
&= \sum_{i\in\mathcal{N}} z^{-i} \underbrace{\sum_{j\in\mathbb{Z}} f_m(jN+i) z^{-jN}}_{\triangleq F_{i,m}(z^N)} \\
&= \sum_{i\in\mathcal{N}} z^{-i} F_{i,m}(z^N) ,
\end{aligned}
\tag{2.16}
$$

and

$$
\begin{aligned}
G_m(z) &\triangleq \mathcal{Z}\{g_m(k)\} \\
&= \sum_{k\in\mathbb{Z}} g_m(k) z^{-k} \\
&= \sum_{j\in\mathbb{Z}} \left[g_m(jN)z^{-jN} + g_m(jN-1)z^{-(jN-1)} + \cdots + g_m(jN-N+1)z^{-(jN-N+1)} \right] \\
&= \sum_{j\in\mathbb{Z}} \sum_{i\in\mathcal{N}} g_m(jN-i) z^{-(jN-i)} \\
&= \sum_{i\in\mathcal{N}} \sum_{j\in\mathbb{Z}} g_m(jN-i) z^{-jN} z^{i} \\
&= \sum_{i\in\mathcal{N}} z^{i} \underbrace{\sum_{j\in\mathbb{Z}} g_m(jN-i) z^{-jN}}_{\triangleq G_{m,i}(z^N)} \\
&= \sum_{i\in\mathcal{N}} z^{i} G_{m,i}(z^N) ,
\end{aligned}
\tag{2.17}
$$

where $m \in \mathcal{M}$, and $F_m(z)$ and $G_m(z)$ are the $\mathcal{Z}$-transforms of $f_m(k)$ and $g_m(k)$, respectively. The transfer functions $F_{i,m}(z)$ are the Type-I polyphase components of order N associated with $F_m(z)$, whereas the transfer functions $G_{m,i}(z)$ are the Type-II polyphase components of order N associated with $G_m(z)$.

Example 2.2 (Polyphase Decomposition) Let us consider the signal $h(n)$ defined in Example 2.1. Determine the Type-I polyphase decomposition of order 3 associated with the transfer function $H(z)$.

Solution. Consider that $i \in \{0, 1, 2\}$. Thus, we have

$$H_i(z) = \sum_{j \in \mathbb{Z}} h(3j+i)z^{-j}, \tag{2.18}$$

yielding

$$\begin{aligned} H_0(z) &= 1 + 8z^{-1} + 64z^{-2}, && (2.19)\\ H_1(z) &= 2 + 16z^{-1} + 128z^{-2}, && (2.20)\\ H_2(z) &= 4 + 32z^{-1}. && (2.21) \end{aligned}$$

Observe that

$$\begin{aligned} H(z) &= H_0(z^3) + z^{-1}H_1(z^3) + z^{-2}H_2(z^3) \\ &= \left(1 + 8z^{-3} + 64z^{-6}\right) + z^{-1}\left(2 + 16z^{-3} + 128z^{-6}\right) + z^{-2}\left(4 + 32z^{-3}\right) \\ &= 1 + 2z^{-1} + 4z^{-2} + 8z^{-3} + 16z^{-4} + 32z^{-5} + 64z^{-6} + 128z^{-7}. \end{aligned} \tag{2.22}$$

Comparing with the solution of Example 2.1, the reader should also notice that $H_{\text{dec}}(z) = H_0(z) = \left[H_0(z^3)\right]_{\downarrow 3}$. This is a useful property that will be further exploited. □

By using a matrix approach, we can rewrite Equations (2.16) and (2.17) as follows:

$$\begin{aligned} \begin{bmatrix} F_0(z) & \cdots F_{M-1}(z) \end{bmatrix} &= \underbrace{\begin{bmatrix} 1 & z^{-1} & \cdots & z^{-(N-1)} \end{bmatrix}}_{\mathbf{d}^T(z)} \underbrace{\begin{bmatrix} F_{0,0}(z^N) & \cdots & F_{0,M-1}(z^N) \\ \vdots & \ddots & \vdots \\ F_{N-1,0}(z^N) & \cdots & F_{N-1,M-1}(z^N) \end{bmatrix}}_{\mathbf{F}(z^N)} \\ &= \mathbf{d}^T(z)\mathbf{F}(z^N), \end{aligned} \tag{2.23}$$

$$\begin{aligned} \begin{bmatrix} G_0(z) \\ \vdots \\ G_{M-1}(z) \end{bmatrix} &= \underbrace{\begin{bmatrix} G_{0,0}(z^N) & \cdots & G_{0,N-1}(z^N) \\ \vdots & \ddots & \vdots \\ G_{M-1,0}(z^N) & \cdots & G_{M-1,N-1}(z^N) \end{bmatrix}}_{\mathbf{G}(z^N)} \underbrace{\begin{bmatrix} 1 \\ \vdots \\ z^{(N-1)} \end{bmatrix}}_{\mathbf{d}(z^{-1})} \\ &= \mathbf{G}(z^N)\mathbf{d}(z^{-1}). \end{aligned} \tag{2.24}$$

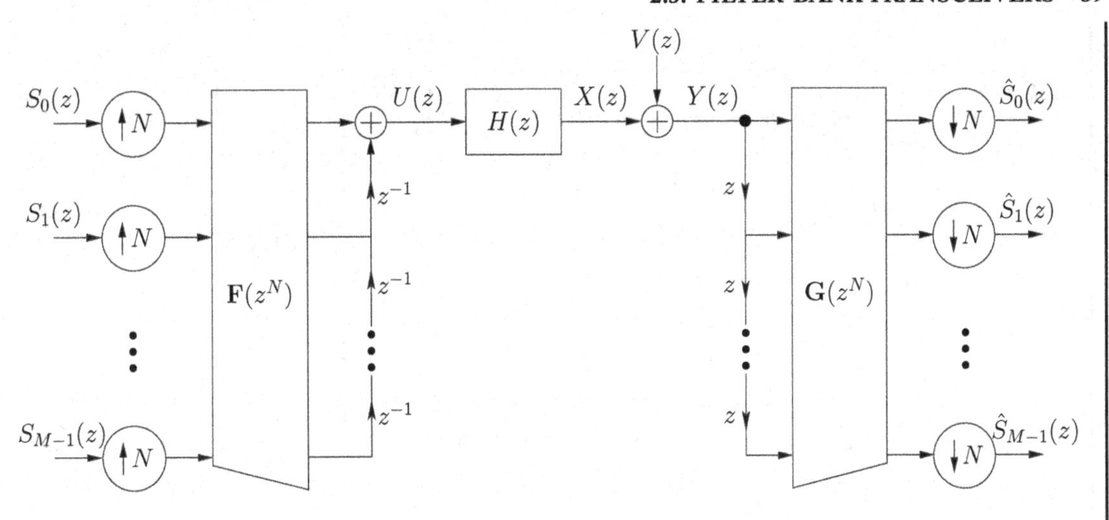

Figure 2.7: Polyphase representation of TMUX systems.

Now, by defining $S_m(z) \triangleq \mathcal{Z}\{s_m(n)\}$, $U(z) \triangleq \mathcal{Z}\{u(k)\}$, $X(z) \triangleq \mathcal{Z}\{x(k)\}$, $V(z) \triangleq \mathcal{Z}\{v(k)\}$, $Y(z) \triangleq \mathcal{Z}\{y(k)\}$, and $\hat{S}_m(z) \triangleq \mathcal{Z}\{\hat{s}_m(n)\}$, then one can write

$$U(z) = \mathbf{d}^T(z)\mathbf{F}(z^N)\underbrace{\begin{bmatrix} S_0(z^N) \\ \vdots \\ S_{M-1}(z^N) \end{bmatrix}}_{\triangleq \mathbf{s}(z)}, \tag{2.25}$$

$$X(z) = H(z)U(z), \tag{2.26}$$

$$Y(z) = X(z) + V(z), \tag{2.27}$$

$$\underbrace{\begin{bmatrix} \hat{S}_0(z) \\ \vdots \\ \hat{S}_{M-1}(z) \end{bmatrix}}_{\triangleq \hat{\mathbf{s}}(z)} = \left[\mathbf{G}(z^N)\mathbf{d}(z^{-1})Y(z)\right]_{\downarrow N}. \tag{2.28}$$

Figure 2.7 illustrates the transceiver model utilizing the polyphase decompositions of the transmitter and receiver subfilters. By employing the noble identities described in Section 2.2, it is possible to transform the transceiver of Figure 2.7 into the equivalent transceiver of Figure 2.8.

The highlighted area of Figure 2.8 that includes delays, forward delays, decimators, interpolators, and the SISO channel model can be represented by a *pseudo-circulant matrix* $\mathbf{H}(z)$ of dimension

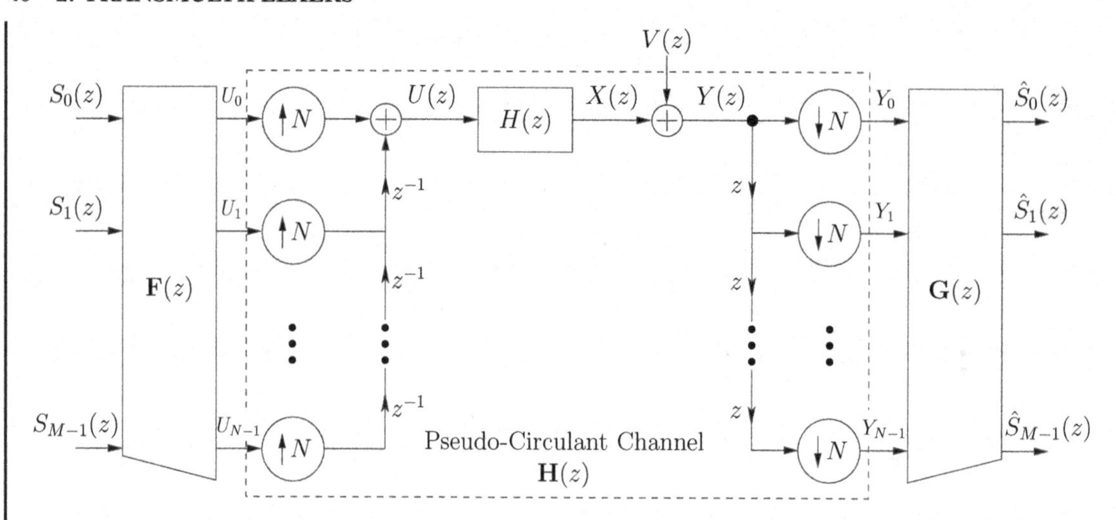

Figure 2.8: Equivalent representation of TMUX systems employing polyphase decompositions.

$N \times N$, given by

$$\mathbf{H}(z) \triangleq \begin{bmatrix} H_0(z) & z^{-1}H_{N-1}(z) & z^{-1}H_{N-2}(z) & \cdots & z^{-1}H_1(z) \\ H_1(z) & H_0(z) & z^{-1}H_{N-1}(z) & \cdots & z^{-1}H_2(z) \\ \vdots & \vdots & \ddots & \vdots & \vdots \\ H_{N-1}(z) & H_{N-2}(z) & H_{N-3}(z) & \cdots & H_0(z) \end{bmatrix}, \tag{2.29}$$

in which

$$H(z) \triangleq \sum_{i \in \mathcal{N}} H_i(z^N)z^{-i} \quad \text{and} \quad H_i(z) \triangleq \sum_{\substack{j \in \mathbb{Z} \\ 0 \leq jN+i \leq L}} h(jN+i)z^{-j}. \tag{2.30}$$

Indeed, given the indexes m and l within the set $\mathcal{N}$, the (m,l)th element of the matrix $\mathbf{H}(z)$, denoted as $[\mathbf{H}(z)]_{ml}$, represents the transfer function from the lth input element of the highlighted area shown in Figure 2.8 to the mth output element of this area. Hence, by assuming that $v(k) = 0$ for all integer number k, if $U_l(z)$ is the lth input at the transmitter end of the highlighted area in Figure 2.8 and $Y_m(z)$ is the mth output of this area at the receiver end, then

$$\begin{aligned} Y_m(z) &= \left[z^{-l}H(z)U_l(z^N)z^m\right]_{\downarrow N} \\ &= \left[z^{m-l}H(z)U_l(z^N)\right]_{\downarrow N} \\ &= U_l(z)\left[z^{m-l}H(z)\right]_{\downarrow N}, \end{aligned} \tag{2.31}$$

in which we have applied the noble identity described in Equation (2.11) and we also have considered that the only non-zero input of the highlighted area in Figure 2.8 is $U_l(z)$.

Therefore, based on Equation (2.31) and on the first type of polyphase representation of the channel-transfer function, we can write

$$
\begin{aligned}
[\mathbf{H}(z)]_{ml} &= \frac{Y_m(z)}{U_l(z)} \\
&= \left[z^{m-l}H(z)\right]_{\downarrow N} \\
&= \left[z^{m-l}\sum_{i\in\mathcal{N}} H_i(z^N)z^{-i}\right]_{\downarrow N} \\
&= \left[\sum_{i\in\mathcal{N}} H_i(z^N)z^{m-l-i}\right]_{\downarrow N} \\
&= \left[H_0(z^N)z^{m-l} + H_1(z^N)z^{m-l-1} + \cdots + H_{N-1}(z^N)z^{m-l-N+1}\right]_{\downarrow N}.
\end{aligned}
\tag{2.32}
$$

We know that the decimation operation retains the first coefficient out of N coefficients within a block, starting from the 0th element. In the $\mathcal{Z}$-domain, this means that the decimation operation keeps only the coefficients which multiply a power of z^N. Thus, the jth coefficient of the decimated signal corresponds to the (jN)th coefficient of the signal before the decimation. Another way of interpreting this fact is that, given an index $i_0 \in \mathcal{N}$ such that $(m-l-i_0)$ is a multiple of N, the decimation operation which appears in expression (2.32) retains the i_0th term $H_{i_0}(z^N)z^{m-l-i_0}$ and decimates it, as illustrated in Example 2.2.

We also know that $-(N-1) \leq m-l \leq N-1$, since m and l are within the set $\mathcal{N}$. Hence, if $m-l \geq 0$, then expression (2.32) yields

$$
[\mathbf{H}(z)]_{ml} = \left[H_{(m-l)}(z^N)\right]_{\downarrow N} = H_{(m-l)}(z). \tag{2.33}
$$

On the other hand, if $m-l < 0$, then

$$
[\mathbf{H}(z)]_{ml} = \left[z^{-N}H_{(N-l+m)}(z^N)\right]_{\downarrow N} = z^{-1}H_{(N-l+m)}(z), \tag{2.34}
$$

confirming the relations described in Equations (2.29) and (2.30).

Figure 2.9 describes the transceiver through the polyphase decomposition of appropriate matrices, including the pseudo-circulant representation of the channel matrix. It is worth noting that the descriptions of Figures 2.6 and 2.9 are equivalent.

As Figure 2.9 illustrates, the transmitted and received vectors are denoted as

$$
\mathbf{s}(n) \triangleq [\, s_0(n) \;\; s_1(n) \;\; \cdots \;\; s_{M-1}(n)\,]^T, \tag{2.35}
$$

$$
\hat{\mathbf{s}}(n) \triangleq [\, \hat{s}_0(n) \;\; \hat{s}_1(n) \;\; \cdots \;\; \hat{s}_{M-1}(n)\,]^T. \tag{2.36}
$$

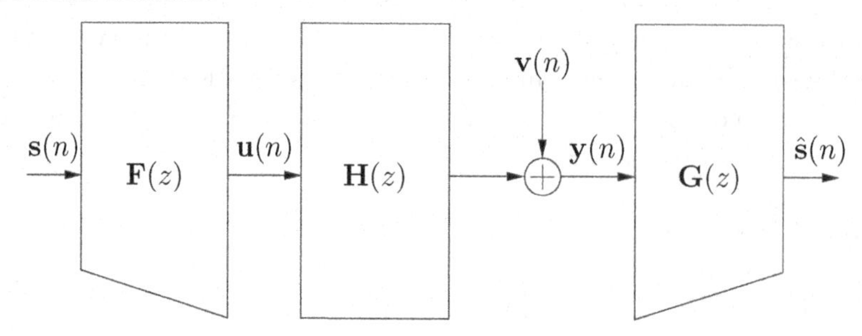

Figure 2.9: Block-based transceivers in $\mathcal{Z}$-domain employing polyphase decompositions.

From Figure 2.9, it is also possible to infer that the transfer matrix $\mathbf{T}(z)$ of the transceiver can be expressed as

$$\mathbf{T}(z) \triangleq \mathbf{G}(z)\mathbf{H}(z)\mathbf{F}(z), \tag{2.37}$$

where we considered the particular case in which $v(k) = 0$ for all integer number k, inspired by the *zero-forcing* (ZF) design. A transceiver is zero forcing whenever $\mathbf{T}(z) = \alpha z^{-d}\mathbf{I}_M$, for some $\alpha \in \mathbb{C}$ and $d \in \mathbb{N}$. Notice that, if there is no noise, a zero-forcing solution is able to retrieve a scaled and delayed version of all transmitted signals.

An important observation about Figure 2.9 is that, in order to be able to recover a block with M transmitted symbols, one must send through the channel at least M elements in a data block, i.e., we must necessarily have $N \geq M$ (this fact explains the shapes of the boxes in Figure 2.9). Nonetheless, if $N = M$ (no redundancy is included), then the matrices $\mathbf{F}(z)$, $\mathbf{H}(z)$, and $\mathbf{G}(z)$ are square matrices and, therefore, a zero-forcing solution would not be achieved using only FIR filters, considering that the channel model is not a simple delay, as explained in [44, 45]. Hence, some redundancy must be introduced in order to work with FIR transceivers.[3]

Now, let us assume that we choose $N > L$, i.e., the interpolation/decimation factor is greater than or equal to the channel order L, a common situation in practice.[4] Based on Equation (2.30), we have that the only integer number j which satisfies the inequality constraint $0 \leq jN + i \leq L$ is $j = 0$, which lead us to conclude that $H_i(z) = h(i)$, for $\mathcal{N} \ni i \leq L$. On the other hand, if there exists $i > L$ within the set $\mathcal{N}$, then $H_i(z) = 0$ since there is no term to be added in order to form $H_i(z)$. In other words, we can say that, for $N > L$, each element of the matrix $\mathbf{H}(z)$ will consist of filters with a single (possibly null) coefficient. In this case, the pseudo-circulant channel matrix in

[3]Employing IIR filters may bring about many drawbacks, such as instability issues. This is the reason why FIR transceivers are the prevalent choice.

[4]Usually, practical block transceivers use $N = M + K$, where K is an integer number larger than or equal to L.

Equation (2.29) is represented by a first-order FIR matrix described as

$$\mathbf{H}(z) = \underbrace{\begin{bmatrix} h(0) & 0 & 0 & \cdots & 0 \\ h(1) & h(0) & 0 & \cdots & 0 \\ \vdots & \vdots & \vdots & \vdots & \vdots \\ h(L) & h(L-1) & \ddots & \cdots & 0 \\ 0 & h(L) & \cdots & \cdots & 0 \\ \vdots & \vdots & \vdots & \vdots & \vdots \\ 0 & 0 & h(L) & \cdots & h(0) \end{bmatrix}}_{\triangleq \mathbf{H}_{\text{ISI}}} + z^{-1} \underbrace{\begin{bmatrix} 0 & \cdots & 0 & h(L) & \cdots & h(1) \\ 0 & 0 & \cdots & 0 & \ddots & \vdots \\ \vdots & \vdots & \vdots & \vdots & \vdots & h(L) \\ 0 & 0 & 0 & \cdots & 0 & 0 \\ 0 & 0 & 0 & \cdots & 0 & 0 \\ \vdots & \vdots & \vdots & \vdots & \vdots & \vdots \\ 0 & 0 & 0 & 0 & \cdots & 0 \end{bmatrix}}_{\triangleq \mathbf{H}_{\text{IBI}}}$$
$$= \mathbf{H}_{\text{ISI}} + z^{-1}\mathbf{H}_{\text{IBI}}. \tag{2.38}$$

Notice that Equation (2.38) implies the following relation in the time domain:

$$\begin{aligned} \mathbf{x}(k) &= \mathbf{H}_{\text{ISI}}\mathbf{u}(k) + \mathbf{H}_{\text{IBI}}\mathbf{u}(k-1) \\ &= \begin{bmatrix} \mathbf{H}_{\text{IBI}} & \mathbf{H}_{\text{ISI}} \end{bmatrix} \begin{bmatrix} \mathbf{u}(k-1) \\ \mathbf{u}(k) \end{bmatrix}, \end{aligned} \tag{2.39}$$

where

$$\mathbf{u}(k) \triangleq [\, u(kN-N+1) \;\; u(kN-N+2) \;\; \cdots \;\; u(kN)\,]^T, \tag{2.40}$$
$$\mathbf{x}(k) \triangleq [\, x(kN-N+1) \;\; x(kN-N+2) \;\; \cdots \;\; x(kN)\,]^T. \tag{2.41}$$

The relationship described in Equation (2.39) makes clear the roles of the matrices $\mathbf{H}_{\text{ISI}}$ and $\mathbf{H}_{\text{IBI}}$. Indeed, matrix $\mathbf{H}_{\text{ISI}}$ mixes the symbols transmitted in the current data block, i.e., such a matrix introduces interferences among the current data-block symbols, while matrix $\mathbf{H}_{\text{IBI}}$ mixes some of the symbols transmitted in the past block. The channel output vector $\mathbf{x}(k)$ is the result of adding both effects: ISI and IBI.

Another way to derive Equation (2.39) is by analyzing what happens in the time domain when a signal $u(k)$ passes through an FIR channel $h(k)$ of order L. In this case, we know that the channel output $x(k)$ is the *linear convolution* between the signals $u(k)$ and $h(k)$, that is, $x(k) = (h * u)(k)$. Hence, if we look at a block of size $N > L$ containing the channel output signals (in other words, if we examine the elements of the vector $\mathbf{x}(k)$), then we can verify that the first L elements of this block are affected by the last L elements of the previous block, due to the channel memory and the way the linear convolution is computed.

It is worth pointing out that generalizations of standard multicarrier communications systems may call for sophisticated transmultiplexer designs in which the transmitted signal is filtered by a precoder with memory consisting of a MIMO FIR filter. The inherent memory at the transmitter can be viewed as a kind of redundancy since a given signal block is transmitted more than once along with neighboring blocks. Sophisticated transmitters may require more complex receivers, but they might allow a reduction in the amount of redundant signals necessary to attain zero-forcing solution, for example. All of these facts indicate that communication engineers should master the TMUX-related tools in order to pursue new advances in communications systems, especially regarding multicarrier transceivers. The case of transceivers with memory will be addressed in Chapter 5. In this chapter, we shall consider the widespread used memoryless systems.

2.4 MEMORYLESS BLOCK-BASED SYSTEMS

The particular and very important case where the transceivers are memoryless, that is, $\mathbf{F}(z) = \mathbf{F}$ and $\mathbf{G}(z) = \mathbf{G}$, is addressed in this section. This case encompasses the *memoryless block-based transceivers*, since these systems do not use data from previous or future blocks in the transmission and reception processing of the current data block. That is, only the current block takes part in the transceiver computations. The traditional OFDM and SC-FD transceivers are well-known examples of memoryless block-based systems.

The non-overlapping behavior associated with memoryless transceivers is only possible if the lengths of the FIR causal subfilters $\{f_m(k)\}_{m\in\mathcal{M}}$ and $\{g_m(k)\}_{m\in\mathcal{M}}$ are less than or equal to N. Indeed, from Equations (2.16) and (2.17), we know that

$$[\mathbf{F}(z)]_{im} = \sum_{j\in\mathbb{Z}} f_m(jN+i)z^{-j}, \tag{2.42}$$

$$[\mathbf{G}(z)]_{mi} = \sum_{j\in\mathbb{Z}} g_m(jN-i)z^{-j}, \tag{2.43}$$

for all pairs of numbers (i, m) within the set $\mathcal{N} \times \mathcal{M}$. Hence, the matrix $\mathbf{F}(z)$ will have memory (i.e., will depend on z) if, and only if, there exists both a non-zero natural[5] number j_0 and a pair of numbers $(i_0, m_0) \in \mathcal{N} \times \mathcal{M}$, such that $f_{m_0}(j_0N + i_0) \neq 0$, which occurs if, and only if, $f_{m_0}(k)$ is a causal impulse response with length larger than N, since $j_0N + i_0 \geq N$. The same conclusion can be drawn for the matrix $\mathbf{G}(z)$.

We shall briefly describe now the main memoryless LTI block transceivers which will be considered throughout this book. Further details will be given in Chapters 3 and 4.

[5]The index j_0 cannot be negative because we are only interested in causal subfilters.

2.4.1 CP-OFDM

The OFDM transceiver employing *cyclic prefix* as redundancy (also known as cyclic-prefix OFDM, or just CP-OFDM) is described by the following transmitter and receiver matrices, respectively:

$$\mathbf{F} \triangleq \underbrace{\begin{bmatrix} \mathbf{0}_{K\times(M-K)} & \mathbf{I}_K \\ \multicolumn{2}{c}{\mathbf{I}_M} \end{bmatrix}}_{\mathbf{A}_{\text{CP}}\in\mathbb{C}^{N\times M}} \mathbf{W}_M^H, \tag{2.44}$$

$$\mathbf{G} \triangleq \mathbf{E}\mathbf{W}_M \underbrace{\begin{bmatrix} \mathbf{0}_{M\times K} & \mathbf{I}_M \end{bmatrix}}_{\mathbf{R}_{\text{CP}}\in\mathbb{C}^{M\times N}}, \tag{2.45}$$

where the integer number K denotes the amount of redundant elements, $\mathbf{W}_M$ is the unitary $M \times M$ discrete Fourier transform (DFT) matrix, that is,

$$[\mathbf{W}_M]_{ml} \triangleq \frac{W_M^{ml}}{\sqrt{M}}, \tag{2.46}$$

with $(m, l) \in \mathcal{M}^2 \triangleq \mathcal{M} \times \mathcal{M}$ and

$$W_M \triangleq \mathrm{e}^{-\mathrm{j}\frac{2\pi}{M}}. \tag{2.47}$$

In addition, $\mathbf{I}_M$ is the $M \times M$ identity matrix, $\mathbf{0}_{M\times N}$ is an $M \times N$ matrix whose entries are zero, and $\mathbf{E}$ is an $M \times M$ diagonal equalizer matrix placed after the removal of the cyclic prefix and the application of the DFT matrix.

As can be noted, the data block to be transmitted has length M, however, due to the prefix, the transceiver actually transmits a block of length $N = M + K$, in which K must be larger than or equal to the channel order L, i.e., one must necessarily have $M \geq K \geq L$ so that the CP-OFDM system works properly. The first K elements are repetitions of the last K elements of the inverse discrete Fourier transform (IDFT) output in order to implement the cyclic prefix.

Matrix $\mathbf{A}_{\text{CP}}$ adds and matrix $\mathbf{R}_{\text{CP}}$ removes the related cyclic prefix. Note that, based on Equation (2.38), the product $\mathbf{R}_{\text{CP}}\mathbf{H}(z)\mathbf{A}_{\text{CP}} \triangleq \mathbf{H}_{\text{c}} \in \mathbb{C}^{M\times M}$ is given by

$$\mathbf{H}_{\text{c}} = \begin{bmatrix} h(0) & 0 & \cdots & 0 & h(L) & \cdots & h(1) \\ h(1) & h(0) & \cdots & 0 & 0 & \ddots & \vdots \\ \vdots & \vdots & \ddots & & & & h(L) \\ h(L) & h(L-1) & & \ddots & \ddots & & 0 \\ 0 & h(L) & & & \ddots & & \vdots \\ \vdots & \ddots & \ddots & & & \ddots & 0 \\ 0 & \cdots & 0 & h(L) & \cdots & & h(0) \end{bmatrix}, \tag{2.48}$$

where we can observe that $\mathbf{R}_{\mathrm{CP}}$ removes the IBI (there is no dependency on z anymore), whereas matrix $\mathbf{A}_{\mathrm{CP}}$ right-multiplies the resulting memoryless matrix $\mathbf{R}_{\mathrm{CP}}\mathbf{H}(z) \in \mathbb{C}^{M \times N}$ so that the overall matrix product is a *circulant matrix* of dimension $M \times M$. Indeed, one can observe that each row of matrix $\mathbf{H}_{\mathrm{c}}$ can be obtained by circular-shifting the related previous row.

After inclusion and removal of the cyclic prefix, the resulting circulant matrix can be diagonalized by its right-multiplication by the IDFT and left-multiplication by the DFT matrices, where these matrices are placed at the transmitter and receiver sides, respectively. Indeed, we have

$$\begin{aligned}\mathbf{W}_M\mathbf{H}_{\mathrm{c}}\mathbf{W}_M^H &= \mathbf{W}_M\begin{bmatrix}\mathbf{h}_0 & \mathbf{h}_1 & \cdots & \mathbf{h}_{M-1}\end{bmatrix}\mathbf{W}_M^H \\ &= \begin{bmatrix}\mathbf{W}_M\mathbf{h}_0 & \mathbf{W}_M\mathbf{h}_1 & \cdots & \mathbf{W}_M\mathbf{h}_{M-1}\end{bmatrix}\mathbf{W}_M^H \\ &= \begin{bmatrix}\widehat{\mathbf{h}}_0 & \widehat{\mathbf{h}}_1 & \cdots & \widehat{\mathbf{h}}_{M-1}\end{bmatrix}\mathbf{W}_M^H,\end{aligned} \tag{2.49}$$

where $\mathbf{h}_m$ is the mth column of matrix $\mathbf{H}_{\mathrm{c}}$ and $\widehat{\mathbf{h}}_m \triangleq \mathbf{W}_M\mathbf{h}_m$ is its DFT. Note that one can interpret the elements of vector $\mathbf{h}_m$ as a periodic discrete-time signal $h_m(k)$ whose period is M, which respects the relation $h_m(k) = h_0(k-m)$, where $h_0(0) = h(0), h_0(1) = h(1), \cdots, h_0(L) = h(L), h_0(L+1) = 0, \cdots, h_0(M-1) = 0, h_0(M) = h(0)$, and so forth. Thus, by remembering the circular-shifting property of the DFT stating that, given

$$\begin{aligned}H_0(l) &\triangleq \mathrm{DFT}\{h_0(k)\} \\ &= \frac{1}{\sqrt{M}}\sum_{k\in\mathcal{M}} h_0(k)W_M^{lk},\end{aligned} \tag{2.50}$$

for all $l \in \mathcal{M}$, then one has

$$\mathrm{DFT}\{h_0(k-m)\} = W_M^{ml}H_0(l). \tag{2.51}$$

Hence, by applying this result, we have

$$\begin{aligned}H_m(l) &= \mathrm{DFT}\{h_m(k)\} \\ &= \mathrm{DFT}\{h_0(k-m)\} \\ &= W_M^{ml}H_0(l),\end{aligned} \tag{2.52}$$

yielding

$$\begin{aligned}\widehat{\mathbf{h}}_m &= \left(\mathrm{diag}\{W_M^{ml}\}_{l\in\mathcal{M}}\right)\widehat{\mathbf{h}}_0 \\ &= \left(\mathrm{diag}\{W_M^{l}\}_{l\in\mathcal{M}}\right)^m\widehat{\mathbf{h}}_0 \\ &= \mathbf{D}^m\widehat{\mathbf{h}}_0,\end{aligned} \tag{2.53}$$

in which $\mathbf{D} \triangleq \text{diag}\{W_M^l\}_{l \in \mathcal{M}}$ denotes an $M \times M$ diagonal matrix whose the (l, l)th element is W_M^l, for each $l \in \mathcal{M}$. We can therefore rewrite Equation (2.49) as

$$\begin{aligned}
\left[\mathbf{W}_M \mathbf{H}_c \mathbf{W}_M^H\right]_{lm} &= \left[\left[\widehat{\mathbf{h}}_0 \;\; \mathbf{D}\widehat{\mathbf{h}}_0 \;\; \mathbf{D}^2\widehat{\mathbf{h}}_0 \;\; \cdots \;\; \mathbf{D}^{M-1}\widehat{\mathbf{h}}_0\right] \mathbf{W}_M^H\right]_{lm} \\
&= \left[H_0(l)W_M^{l0} \quad H_0(l)W_M^{l1} \quad \cdots \quad H_0(l)W_M^{l(M-1)}\right] \begin{bmatrix} W_M^{-m.0} \\ W_M^{-m.1} \\ \vdots \\ W_M^{-m(M-1)} \end{bmatrix} / \sqrt{M} \\
&= \frac{H_0(l)}{\sqrt{M}} \sum_{i \in \mathcal{M}} W_M^{li} W_M^{-im} \\
&= \frac{H_0(l)}{\sqrt{M}} \sum_{i \in \mathcal{M}} W_M^{i(l-m)}.
\end{aligned} \tag{2.54}$$

If $l = m$, then $W_M^{i(l-m)} = 1$ for all $i \in \mathcal{M}$, implying

$$\left[\mathbf{W}_M \mathbf{H}_c \mathbf{W}_M^H\right]_{ll} = \frac{H_0(l)}{\sqrt{M}} \times M = \sqrt{M} H_0(l), \tag{2.55}$$

while, if $l \neq m$, then $W_M^{(l-m)} \neq 1$, implying

$$\begin{aligned}
\left[\mathbf{W}_M \mathbf{H}_c \mathbf{W}_M^H\right]_{lm} &= \frac{H_0(l)}{\sqrt{M}} \left(\frac{W_M^{M(l-m)} - 1}{W_M^{(l-m)} - 1}\right) \\
&= \frac{H_0(l)}{\sqrt{M}} \left(\frac{1 - 1}{W_M^{(l-m)} - 1}\right) \\
&= 0.
\end{aligned} \tag{2.56}$$

Therefore, we can conclude that

$$\begin{aligned}
\mathbf{\Lambda} &\triangleq \mathbf{W}_M \mathbf{H}_c \mathbf{W}_M^H \\
&= \mathbf{W}_M \mathbf{R}_{\text{CP}} \mathbf{H}(z) \mathbf{A}_{\text{CP}} \mathbf{W}_M^H \\
&= \text{diag}\left\{\sqrt{M}\widehat{\mathbf{h}}_0\right\} \\
&= \text{diag}\left\{\sqrt{M}\mathbf{W}_M \mathbf{h}_0\right\} \\
&= \text{diag}\left\{\sqrt{M}\mathbf{W}_M \begin{bmatrix} \mathbf{h} \\ \mathbf{0}_{(M-L-1)\times 1} \end{bmatrix}\right\} \\
&\triangleq \text{diag}\{\lambda_m\}_{m \in \mathcal{M}},
\end{aligned} \tag{2.57}$$

in which $\mathbf{h} \triangleq [\, h(0) \;\; h(1) \;\; \cdots \;\; h(L)\,]^T$.

Matrix $\mathbf{\Lambda}$ includes at its diagonal the distortion imposed by the channel on each symbol of the data block. Hence, the model of a CP-OFDM transceiver is described by

$$\hat{\mathbf{s}} \triangleq \mathbf{E}\mathbf{\Lambda}\mathbf{s} + \mathbf{E}\mathbf{v}', \tag{2.58}$$

with $\mathbf{v}' \triangleq \mathbf{W}_M \mathbf{R}_{\text{CP}}\mathbf{v}$ and, for the sake of simplicity, the time dependency of the expressions was omitted. As can be observed, the estimates of the transmitted symbols are uncoupled, that is, each symbol can be estimated independently of any other symbol within the related block, avoiding intersymbol interference. One can interpret this fact as if each symbol were transmitted through a flat-fading subchannel.

From a signal processing perspective, the model described in Equation (2.58) has a simple interpretation. Indeed, the addition and removal of the cyclic prefix turns the linear convolution described in Equation (2.39) into a *circular convolution*. In this case, the CP-OFDM system loads each subcarrier in the frequency domain with a constellation symbol and, after that, performs the inverse discrete Fourier transformation, generating a vector in the time domain. The elements of this vector can be thought as a periodic signal which is processed by the channel through a linear convolution. After that, the signal is brought back to the original frequency domain. A basic fact of digital signal processing is that the circular convolution of two signals can be implemented in the frequency domain by performing the product of the DFTs of the related signals. Therefore, the CP-OFDM system can be further simplified if we take this fact into account. All we have to do is to perform the entire processing in the frequency domain. The symbols which are loaded at each subcarrier can be directly mapped to the received signals at each subcarrier by performing the product with the frequency response of the channel (DFT of the zero-padded impulse response, as in Equation (2.57)).

The equalizer $\mathbf{E}$ for this transceiver can be defined in several ways, where the most popular are the *zero-forcing* (ZF) and the *minimum mean square error* (MMSE) equalizers. In the ZF solution, it is aimed to undo the distortions introduced by the channel. Indeed, when there is no noise, the ZF solution is able to perfectly recover the transmitted vector. It is assumed that matrix $\mathbf{\Lambda}$ can be inverted, thus yielding

$$\mathbf{E}_{\text{ZF}} \triangleq \mathbf{\Lambda}^{-1}. \tag{2.59}$$

As for the MMSE solution, there is no requirement that matrix $\mathbf{\Lambda}$ be invertible since this latter operation is not needed. In fact, the linear MMSE equalizer matrix is the solution to the following optimization problem:

$$\mathbf{E}_{\text{MMSE}} \triangleq \arg\left\{\min_{\forall \mathbf{E} \in \mathbb{C}^{M \times M}} \mathcal{J}(\mathbf{E})\right\}, \tag{2.60}$$

where $\mathcal{J}$ is a real-valued function of a complex-valued matrix argument defined as

$$\begin{aligned}\mathcal{J}(\mathbf{E}) &\triangleq E\left[\|\mathbf{s}-\mathbf{E}(\mathbf{\Lambda s}+\mathbf{v}')\|_2^2\right]\\ &= E\left[\left(\mathbf{s}-\mathbf{E\Lambda s}-\mathbf{Ev}'\right)^H\left(\mathbf{s}-\mathbf{E\Lambda s}-\mathbf{Ev}'\right)\right]\\ &= \operatorname{tr}\left\{E\left[\left(\mathbf{s}-\mathbf{E\Lambda s}-\mathbf{Ev}'\right)\left(\mathbf{s}-\mathbf{E\Lambda s}-\mathbf{Ev}'\right)^H\right]\right\}\\ &= \operatorname{tr}\left\{\sigma_s^2\mathbf{I}_M+\sigma_s^2\mathbf{E\Lambda\Lambda}^H\mathbf{E}^H-\sigma_s^2\mathbf{E\Lambda}-\sigma_s^2\mathbf{\Lambda}^H\mathbf{E}^H+\sigma_v^2\mathbf{EE}^H\right\},\end{aligned} \tag{2.61}$$

where $E[\cdot]$ and $\operatorname{tr}\{\cdot\}$ are the expected value and trace operators, respectively. The derivation above assumes that the transmitted symbols and environment noise within a block are independent and identically distributed (i.i.d.), originating from a wide-sense stationary (WSS) white random sequences with zero means and uncorrelated. These assumptions imply that $E[\mathbf{sv}'^H] = E[\mathbf{s}]E[\mathbf{v}']^H = \mathbf{0}_{M\times M} = E[\mathbf{v}']E[\mathbf{s}]^H = E[\mathbf{v}'\mathbf{s}^H]$ and that $E[\mathbf{ss}^H] = \sigma_s^2\mathbf{I}_M$ and $E[\mathbf{v}'\mathbf{v}'^H] = \sigma_v^2\mathbf{I}_M$, where the positive real numbers σ_s^2 and σ_v^2 are the variances of the related WSS random sequences.[6]

Now, by using the following derivatives of scalar functions of complex matrices [83]:

$$\frac{\partial \operatorname{tr}\{\mathbf{ZAZ}^H\}}{\partial \mathbf{Z}^*} = \mathbf{ZA}, \tag{2.62}$$

$$\frac{\partial \operatorname{tr}\{\mathbf{AZ}^H\}}{\partial \mathbf{Z}^*} = \mathbf{A}, \tag{2.63}$$

then we have

$$\frac{\partial \mathcal{J}(\mathbf{E})}{\partial \mathbf{E}^*} = \sigma_s^2\mathbf{E\Lambda\Lambda}^H - \sigma_s^2\mathbf{\Lambda}^H + \sigma_v^2\mathbf{E}. \tag{2.64}$$

We know that the optimal solution $\mathbf{E}_{\text{MMSE}}$ is such that $\frac{\partial\mathcal{J}(\mathbf{E}_{\text{MMSE}})}{\partial\mathbf{E}^*} = \mathbf{0}_{M\times M}$, which implies that[7]

$$\begin{aligned}\mathbf{E}_{\text{MMSE}} &= \mathbf{\Lambda}^H\left(\mathbf{\Lambda\Lambda}^H+\frac{\sigma_v^2}{\sigma_s^2}\mathbf{I}_M\right)^{-1}\\ &= \operatorname{diag}\left\{\frac{\lambda_m^*}{|\lambda_m|^2+\frac{\sigma_v^2}{\sigma_s^2}}\right\}_{m\in\mathcal{M}}.\end{aligned} \tag{2.65}$$

It is worth highlighting that the CP-OFDM transceiver is the most popular type of OFDM-based techniques which are employed in practical applications.

2.4.2 ZP-OFDM

An alternative OFDM system inserts zeros as redundancy and is called zero-padding OFDM (ZP-OFDM). There are many variants of ZP-OFDM. One possible choice is the ZP-OFDM-OLA

[6] In this book, we shall not employ distinct notations for deterministic and random variables.
[7] We encourage the reader to justify why this is actually the minimum solution of the objective function.

(overlap-and-add) whose transmitter and receiver matrices are implemented as

$$\mathbf{F} \triangleq \underbrace{\begin{bmatrix} \mathbf{I}_M \\ \mathbf{0}_{K\times M} \end{bmatrix}}_{\mathbf{A}_{\mathrm{ZP}} \in \mathbb{C}^{N\times M}} \mathbf{W}_M^H, \tag{2.66}$$

$$\mathbf{G} \triangleq \mathbf{E}\mathbf{W}_M \underbrace{\begin{bmatrix} \mathbf{I}_M & \begin{matrix} \mathbf{I}_K \\ \mathbf{0}_{(M-K)\times K} \end{matrix} \end{bmatrix}}_{\mathbf{R}_{\mathrm{ZP}} \in \mathbb{C}^{M\times N}}, \tag{2.67}$$

where, as in the CP-OFDM case, $K \geq L$ elements are inserted as redundancy, and $N = M + K$.

The name OLA stems from the way the received signals are processed by the matrix $\mathbf{R}_{\mathrm{ZP}}$. Matrices $\mathbf{A}_{\mathrm{ZP}}$ and $\mathbf{R}_{\mathrm{ZP}}$ perform the addition and removal of the guard period of zero redundancy, respectively. The matrix product $\mathbf{R}_{\mathrm{ZP}}\mathbf{H}(z)\mathbf{A}_{\mathrm{ZP}} \in \mathbb{C}^{M\times M}$ is given by

$$\mathbf{R}_{\mathrm{ZP}}\mathbf{H}(z)\mathbf{A}_{\mathrm{ZP}} = \begin{bmatrix} h(0) & 0 & \cdots & 0 & h(L) & \cdots & h(1) \\ h(1) & h(0) & \cdots & 0 & 0 & \ddots & \vdots \\ \vdots & \vdots & \ddots & & & & h(L) \\ h(L) & h(L-1) & & \ddots & \ddots & & 0 \\ 0 & h(L) & & & \ddots & & \vdots \\ \vdots & \ddots & \ddots & & & \ddots & 0 \\ 0 & \cdots & 0 & h(L) & \cdots & & h(0) \end{bmatrix} = \mathbf{R}_{\mathrm{CP}}\mathbf{H}(z)\mathbf{A}_{\mathrm{CP}}. \tag{2.68}$$

As can be observed, matrix $\mathbf{A}_{\mathrm{ZP}}$ removes the interblock interference, whereas matrix $\mathbf{R}_{\mathrm{ZP}}$ left-multiplies the resulting memoryless *Toeplitz matrix*[8] $\mathbf{H}(z)\mathbf{A}_{\mathrm{ZP}} \in \mathbb{C}^{N\times M}$ so that the overall product becomes a circulant matrix of dimension $M \times M$. The reader should note that $\mathbf{R}_{\mathrm{ZP}}\mathbf{H}(z)\mathbf{A}_{\mathrm{ZP}} = \mathbf{R}_{\mathrm{CP}}\mathbf{H}(z)\mathbf{A}_{\mathrm{CP}} = \mathbf{H}_c$.

The ZP-OFDM-OLA transceiver discussed here is a simplified version of a more general transceiver proposed in [55].[9] In fact, the general transceiver allows the recovery of the transmitted symbols using zero-forcing equalizers independently of the locations of the channel zeros, unlike the ZP-OFDM-OLA or CP-OFDM that might have zero eigenvalues under certain channel conditions. Unfortunately, from the computational point of view, this transceiver implementation is not as simple as, for instance, the CP-OFDM, since the equivalent channel matrix is not circulant, turning impossible its diagonalization through fast transforms, such as fast Fourier transform (FFT).[10] Furthermore, even for the design of a simple ZF equalizer, the general ZP-OFDM transceiver

[8] See Subsection 4.3.2 for a formal definition of a Toeplitz matrix.
[9] There are other variants of ZP-OFDM, such as the ZP-OFDM-FAST.
[10] Actually, it is possible to implement the general ZP-OFDM system using FFTs, but without diagonalizing the equivalent channel matrix (see Chapter 4).

would require the inverse of a Toeplitz matrix, being therefore more complex than the inversion of a circulant matrix required by a ZP-OFDM-OLA system.

2.4.3 CP-SC-FD

The cyclic-prefix single-carrier with frequency-domain (CP-SC-FD) equalization transceiver employs cyclic prefix as redundancy and it is closely related to the CP-OFDM transceiver. The CP-SC-FD system is described by the following transmitter and receiver matrices:

$$\mathbf{F} \triangleq \begin{bmatrix} \mathbf{0}_{K\times(M-K)} & \mathbf{I}_K \\ \multicolumn{2}{c}{\mathbf{I}_M} \end{bmatrix}, \tag{2.69}$$

$$\mathbf{G} \triangleq \mathbf{W}_M^H \mathbf{E} \mathbf{W}_M \begin{bmatrix} \mathbf{0}_{M\times K} & \mathbf{I}_M \end{bmatrix}. \tag{2.70}$$

2.4.4 ZP-SC-FD

The zero-padding single-carrier with frequency-domain (ZP-SC-FD) equalization transceiver inserts redundant zeros to the block to be transmitted, as in the ZP-OFDM transceiver. The ZP-SC-FD-OLA version may be modeled through the following transmitter and receiver matrices:

$$\mathbf{F} \triangleq \begin{bmatrix} \mathbf{I}_M \\ \mathbf{0}_{K\times M} \end{bmatrix}, \tag{2.71}$$

$$\mathbf{G} \triangleq \mathbf{W}_M^H \mathbf{E} \mathbf{W}_M \begin{bmatrix} \mathbf{I}_M & \begin{matrix} \mathbf{I}_K \\ \mathbf{0}_{(M-K)\times K} \end{matrix} \end{bmatrix}. \tag{2.72}$$

2.4.5 ZP-ZJ TRANSCEIVERS

Lin and Phoong [39, 40, 44] showed that the amount of redundancy (guard samples) $K \triangleq N - M \in \mathbb{N}$ required to eliminate IBI and ISI in memoryless block-based transceivers must satisfy the inequality $2K \geq L$. They proposed a family of memoryless discrete multitone transceivers with reduced redundancy. A particular transceiver of interest to our studies here is the *zero-padding zero-jamming* (ZP-ZJ) system, which is characterized by the following transmitter and receiver matrices:

$$\mathbf{F} \triangleq \begin{bmatrix} \overline{\mathbf{F}} \\ \mathbf{0}_{K\times M} \end{bmatrix}_{N\times M}, \tag{2.73}$$

$$\mathbf{G} \triangleq \begin{bmatrix} \mathbf{0}_{M\times(L-K)} & \overline{\mathbf{G}} \end{bmatrix}_{M\times N}, \tag{2.74}$$

in which $\overline{\mathbf{F}} \in \mathbb{C}^{M\times M}$ and $\overline{\mathbf{G}} \in \mathbb{C}^{M\times(M+2K-L)}$.

The transfer matrix related to this transceiver is given by

$$\mathbf{T}(z) = \mathbf{G}\mathbf{H}(z)\mathbf{F} = \overline{\mathbf{G}}\,\overline{\mathbf{H}}\,\overline{\mathbf{F}} = \overline{\mathbf{T}}, \tag{2.75}$$

where, after removing the redundancy, the *effective channel matrix* is defined as

$$\overline{\mathbf{H}} \triangleq \begin{bmatrix} h(L-K) & \cdots & h(0) & 0 & 0 & \cdots & 0 \\ \vdots & \ddots & & & & & \vdots \\ h(K) & \ddots & & & & & 0 \\ \vdots & \ddots & & & \ddots & & h(0) \\ h(L) & & & & & & \vdots \\ 0 & & & & \ddots & & h(L-K) \\ \vdots & & & & & & \vdots \\ 0 & \cdots & 0 & 0 & h(L) & \cdots & h(K) \end{bmatrix} \in \mathbb{C}^{(M+2K-L)\times M}. \tag{2.76}$$

Considering $v(k) = 0$, for all $k \in \mathbb{Z}$, we have

$$\hat{\mathbf{s}}(n) = \overline{\mathbf{G}}\,\overline{\mathbf{H}}\,\overline{\mathbf{F}}\mathbf{s}(n) = \overline{\mathbf{T}}\mathbf{s}(n). \tag{2.77}$$

Observe that the requirement of having $2K - L \geq 0$ makes sense when we analyze the above expression. Indeed, in order to recover the M transmitted symbols, the memoryless transfer matrix $\overline{\mathbf{T}}$ of dimension $M \times M$ must be full-rank. This means that $\min\{M, M + 2K - L\} \geq M$, i.e., $2K - L \geq 0 \Leftrightarrow K \geq \lceil \frac{L}{2} \rceil$.

For this transceiver there are some constraints to be imposed upon the channel impulse response model so that a zero-forcing solution exists. These constraints are related to the concept of congruous zeros.[11] The *congruous zeros* of a transfer function $H(z)$ are the distinct zeros $z_0, z_1, \cdots, z_{\mu-1} \in \mathbb{C}$, with $\mu \in \mathbb{N}$, which meet the following condition: $z_i^N = z_j^N$, with $H(z_i) = H(z_j) = 0$, for all $i, j \in \{0, 1, \cdots, \mu - 1\}$. Note that μ is a function of N. As shown in [44], the channel model must satisfy the constraint $\mu(N) \leq K$, where $\mu(N)$ denotes the cardinality (number of elements) of the largest set of congruous zeros with respect to N.

Therefore, assuming the existence of minimum-redundancy solutions for a given channel, i.e., considering that $\mu(N) \leq L/2 \in \mathbb{N}$, then the ZF solution is such that its associated receiver matrix is given by

$$\overline{\mathbf{G}} \triangleq (\overline{\mathbf{H}}\,\overline{\mathbf{F}})^{-1} = \overline{\mathbf{F}}^{-1}\overline{\mathbf{H}}^{-1}, \tag{2.78}$$

where $\overline{\mathbf{H}} \in \mathbb{C}^{M\times M}$ is given and $\overline{\mathbf{F}}$ is predefined.

This solution is computationally intensive since it requires the inversions of $M \times M$ matrices, entailing $\mathcal{O}(M^3)$ arithmetic operations. The conventional OFDM and SC-FD transceivers need $\mathcal{O}(M \log M)$ operations for the design of ZF and MMSE equalizers. The equalization process associated with the minimum-redundancy solution consists of multiplying the received vector by the receiver matrix, entailing $\mathcal{O}(M^2)$ operations. This complexity is high as compared to that of

[11]We shall address this topic in Chapter 5 in a more detailed manner.

$\mathcal{O}(M \log M)$ required by traditional OFDM and SC-FD transceivers. This efficient equalization originates from the use of DFT matrices as well as the multiplication by memoryless diagonal matrices, as explained in this chapter.

2.5 CONCLUDING REMARKS

This chapter has briefly reviewed the modeling of communications systems using the transmultiplexer framework. The LTI memoryless transceivers were the main focus of our presentation. Among these transceivers we particularly addressed the CP-OFDM, ZP-OFDM-OLA, CP-SC-FD, and ZP-SC-FD-OLA transceivers, highlighting their corresponding ZF and MMSE designs. Some results taken from the open literature related to transceivers with reduced redundancy (ZP-ZJ systems) were also discussed.

A lesson learned from this chapter is that the conventional OFDM and SC-FD transceivers are very efficient since the receiver and the equalizer have very simple implementations. These systems capitalize on the circulant structure of the effective channel matrix whenever a cyclic prefix of length at least L is inserted, where L is the channel order. The circulant matrices can be diagonalized using a pair of DFT and IDFT transformations. Chapter 3 contains an in-depth description of OFDM and SC-FD techniques, including details about the effects of employing different types of prefixes/suffixes.

A further query is if it is possible to derive transceivers similar to the OFDM and SC-FD while employing reduced redundancy and whose implementations rely on fast transforms as well. The answer to such a query is yes, as will be clarified later on in Chapter 4. Another relevant question is if it is possible to reduce even more the transmitted redundancy by working with time-varying transmultiplexers with memory. Once again, the answer is yes, as described in Chapter 5.

CHAPTER 3

OFDM

3.1 INTRODUCTION

As discussed in the previous chapters, the *orthogonal frequency-division multiplexing* (OFDM) is a transmission technique that is currently used in a number of wired and wireless systems. This chapter describes OFDM in more detail, starting from its original conception in the continuous-time domain (herein called, *analog OFDM*) and arriving at its current implementation in the discrete-time domain. In fact, the discrete-time description of OFDM has already been addressed in Section 2.4 of Chapter 2. However, that description is solely based on the useful mathematical properties related to circulant matrices, without necessarily calling for physical intuition of actual transmissions. The focus of the present chapter, on the other hand, is to motivate the construction of the OFDM system by analyzing its very insightful analog version and to derive the discrete-time implementation from this physically meaningful continuous-time system. Indeed, it was only with the widespread use of digital integrated circuit technology that the *discrete-time OFDM* transmission technique became popular, especially due to the existence of fast Fourier transform (FFT) algorithms, which enable efficient computations of the discrete Fourier transform (DFT) employed for modulation.

From a historical perspective, the origins of *frequency-division multiplexing* (FDM) date back to the late nineteen century, according to the review article by S. B. Weinstein [88]. The analog version of OFDM was first proposed by R. W. Chang in 1966 [10], who filed a patent that was granted in 1970 [11]. A major breakthrough was the perception that the use of analog subcarrier oscillators and their corresponding coherent demodulators could be avoided by replacing them by DFT-based transceivers. In this context, S. B. Weinstein and P. M. Ebert [89] were the originators of the DFT-based modulation and demodulation schemes. Another key result related to the digital OFDM implementation was conceived by A. Peled and A. Ruiz [64] who advanced the use of cyclic prefix as solution for maintaining orthogonality among subcarriers at the receiver side. Although the analog and digital versions of OFDM systems are closely related, they are not always fully equivalent as discussed in [43]. OFDM has become widely adopted in commercial applications, thus explaining why there are so many works addressing its history [13, 88].

This chapter is organized as follows. Section 3.2 describes the origins of OFDM in its analog version. Such topic is particularly interesting for understanding the choices of some important parameters, such as the OFDM symbol duration, sampling period, and guard period. In addition, Section 3.2 also introduces the importance of *orthogonality* in OFDM. Section 3.3 describes the discrete-time implementation of OFDM systems. The idea of Section 3.3 is to connect what we have seen in Section 2.4 of Chapter 2 with the theory of analog OFDM. Section 3.4 describes some variants of OFDM-based systems, including *single-carrier with frequency-domain* equalization (SC-FD), *zero-padding* (ZP) schemes, *coded OFDM* (C-OFDM), and *discrete multitone* (DMT) systems. Finally, some conclusions are drawn in Section 3.5.

3.2 ANALOG OFDM

Digital communications require the conversion of a discrete-time signal to a continuous-time signal that is actually transmitted. Such an operation is performed by a *digital-to-analog converter* (DAC). If we assume that s_m denotes the mth element of a discrete-time signal, with $m \in \mathcal{M} \triangleq \{0, 1, \ldots, M-1\}$ and $M \in \mathbb{N}$, then the conversion to its related *continuous-time baseband signal* $s_{\text{DAC}}(t)$ can be (theoretically) implemented by first multiplying each element s_m by a continuous-time Dirac impulse $\delta(t - mT)$, and then passing the resulting signal through a linear time-invariant (LTI) analog filter with impulse response $p(t)$. In this context, the positive real-valued parameter T denotes the sampling period of the DAC. Mathematically, we have

$$\begin{aligned} s_{\text{DAC}}(t) &\triangleq \left[\sum_{m\in\mathcal{M}} s_m \delta(t - mT)\right] * p(t) \\ &= \sum_{m\in\mathcal{M}} s_m \left[\delta(t - mT) * p(t)\right] \\ &= \sum_{m\in\mathcal{M}} s_m p(t - mT), \end{aligned} \tag{3.1}$$

where $*$ represents linear convolution.

In other words, the usual digital-to-analog conversion, which is always present in digital communications, can be regarded as a time-division multiplexing (TDM) operation of the elements which compose a discrete-time signal. A natural question arises at this point: is there anything we can do to perform this conversion in a frequency-division multiplexing (FDM) manner? The answer is yes, and we will show that such an FDM-based representation is a natural starting point to conceive the so-called analog OFDM.

3.2.1 FROM TDM TO FDM

In general, the continuous-time signal $s(t)$ associated with a discrete-time signal s_m can be described as

$$s(t) \triangleq \sum_{m\in\mathcal{M}} s_m p_m(t), \tag{3.2}$$

where $p_m(t)$ is a continuous-time pulse signal. The choice of the pulse signal determines how the elements of the discrete-time signal are distributed over the time-frequency plane.

For example, by choosing $p_m(t)$ in a TDM fashion, so that $p_m(t) \triangleq p(t - mT)$ whose time support is the real interval $[mT, (m+1)T)$, we generate the following continuous-time signal (see

Equation (3.1) as well):

$$s_{\text{TDM}}(t) \triangleq \sum_{m\in\mathcal{M}} s_m p(t - mT). \tag{3.3}$$

The former equation implies that $s_{\text{TDM}}(t)$ is a concatenation of pulses $p(t)$, each of them starting at time $t = mT$ with duration of T seconds, modulated by their corresponding symbol s_m originating from a digital modulator. From Fourier analysis we know that the Fourier transform (FT) of $s_{\text{TDM}}(t)$ is

$$S_{\text{TDM}}(\omega) = \sum_{m\in\mathcal{M}} s_m P(\omega)\mathrm{e}^{-\mathrm{j}\omega mT}, \tag{3.4}$$

where the FT of $p(t)$ is represented by $P(\omega)$, whose bandwidth is Ω. Since in TDM schemes each symbol is transmitted in a time slot with T seconds of duration, then the transmission of M symbols lasts MT seconds. In frequency domain each of these symbols occupies the entire available bandwidth Ω.

On the other hand, in FDM schemes we utilize a dual strategy for signal transmission. Indeed, in FDM each symbol occupies a portion of the whole channel bandwidth Ω. The frequency response of the transmitted signal is

$$S_{\text{FDM}}(\omega) \triangleq \sum_{m\in\mathcal{M}} s_m P'(\omega - m\Omega'), \tag{3.5}$$

in which the support of $P'(\omega - m\Omega')$ is the real interval $[m\Omega', (m+1)\Omega')$, where Ω' represents a fraction of the channel bandwidth Ω, that is, $\Omega = M\Omega'$. A signal with such representation in the frequency domain can be written in time domain as

$$\begin{aligned} s_{\text{FDM}}(t) &= \sum_{m\in\mathcal{M}} s_m p'(t)\mathrm{e}^{\mathrm{j}m\Omega' t} \\ &= \left(\sum_{m\in\mathcal{M}} s_m \mathrm{e}^{\mathrm{j}m\Omega' t}\right) p'(t). \end{aligned} \tag{3.6}$$

Thus, FDM transmission is obtained when we choose $p_m(t) \triangleq p'(t)\mathrm{e}^{\mathrm{j}m\Omega' t}$, which implies that the symbols s_m are all superposed in time domain.

Equation (3.6) will be key to the forthcoming description of analog OFDM since it reveals clearly the central role that complex exponentials play in FDM-based transmissions. Note that the existence of such complex exponentials is a natural consequence of the FDM characteristic. In addition, Equation (3.5) is related to the ideal concept of multicarrier systems, which focus on dividing the available channel into many narrowband subchannels so that the channel frequency response can be considered constant in each individual subchannel. In the remaining of this section it will be shown how analog OFDM exploits FDM transmissions in an efficient manner.

3.2.2 ORTHOGONALITY AMONG SUBCARRIERS

The starting point of analog OFDM is the term between parenthesis in expression (3.6), where each entry (symbol) of a data block modulates a subcarrier, which can be interpreted as a tone. In order for these symbols to be easily recovered at the receiver, the subcarriers should be orthogonal. The concept of orthogonality among subcarriers will be explored in the following discussion.

Let us consider that the OFDM subcarriers consist of equally spaced tones in frequency domain. Indeed, if we define $f_m, f_{m+1} \in \mathbb{R}$ as the central frequencies corresponding to the mth and $(m+1)$th subcarriers, respectively, where

$$m \in \mathcal{M} \triangleq \{0, 1, \ldots, M-1\} \tag{3.7}$$

is the subcarrier index, and M is a positive integer number representing the number of subcarriers, then the frequency separation between two consecutive subcarriers is

$$\frac{1}{\Delta} \triangleq f_{m+1} - f_m, \tag{3.8}$$

for all $m \in \mathcal{M} \setminus \{M-1\}$. Note that, by assuming that a subcarrier is comprised of a single tone, let us say f_m, the time-domain representation of such subcarrier consists of a complex exponential at that frequency,[1] that is, $e^{j2\pi f_m t}$.

The transmission of a block with M symbols belonging to a given constellation $\mathcal{C} \subset \mathbb{C}$, in which each symbol is denoted as $s_m(n) \in \mathcal{C}$, is performed by transmitting these symbols using subcarriers with distinct central frequencies. In this context, n is an integer number that identifies the block with M constellation symbols. Such association between symbols and subcarriers is exemplified below as

$$\begin{aligned} s_0(n) &\mapsto f_0 \triangleq 0, \\ s_1(n) &\mapsto f_1 = \frac{1}{\Delta}, \\ &\vdots \\ s_{M-1}(n) &\mapsto f_{M-1} = \frac{M-1}{\Delta}. \end{aligned} \tag{3.9}$$

Hence, the mth symbol is associated with the subcarrier whose central frequency is $f_m = \frac{m}{\Delta}$, for each m within $\mathcal{M}$. The nth data block to be transmitted, usually called *OFDM symbol*,[2] is a complex signal denoted by $\hat{u}_n(t)$, in which t is a real variable representing time. The OFDM symbol $\hat{u}_n(t)$ is generated as the superposition of the subcarriers, each of them modulated by its

[1]Indeed, a single-tone signal whose tone is f_m is a signal whose frequency-domain representation consists of an impulse centered at frequency f_m. From Fourier analysis, we know that the inverse Fourier transform of an impulse at frequency f_m corresponds to $e^{j2\pi f_m t}$.

[2]The reader should not confuse the terms *symbol* and *OFDM symbol*. While the former is a complex number generated at the output of a digital modulator (see Section 1.2), the latter is associated with a collection of constellation symbols.

corresponding symbol $s_m(n)$, yielding

$$\begin{aligned}\hat{u}_n(t) &\triangleq \frac{1}{\sqrt{\hat{T}}}\sum_{m\in\mathcal{M}} s_m(n)\mathrm{e}^{\mathrm{j}2\pi f_m(t-n\hat{T})}\\ &= \frac{1}{\sqrt{\hat{T}}}\sum_{m\in\mathcal{M}} s_m(n)\mathrm{e}^{\mathrm{j}\frac{2\pi}{\Delta}m(t-n\hat{T})}, \quad \text{for } n\hat{T} \le t < (n+1)\hat{T},\end{aligned} \tag{3.10}$$

where $\hat{T}$ is a positive real number representing the duration of an OFDM symbol.

In the time domain, if we assume that each symbol $s_m(n)$ represents an entry of a serial data, then the time support of each symbol is $\frac{\hat{T}}{M}$, since each OFDM symbol is comprised of M constellation symbols, as described in Equation (3.10). The OFDM symbol duration $\hat{T}$ must be long enough to keep the subcarriers orthogonal to each other so that the individual data symbols can be extracted from the OFDM symbol.

Indeed, for any two modulated subcarriers, for example $s_m(n)\mathrm{e}^{\mathrm{j}\frac{2\pi}{\Delta}mt}$ and $s_{m+l}(n)\mathrm{e}^{\mathrm{j}\frac{2\pi}{\Delta}(m+l)t}$, where l is an integer number such that $m+l \in \mathcal{M}$, their *temporal cross-correlation* computed over the OFDM symbol duration is given by

$$\begin{aligned}&\frac{1}{\hat{T}}\int_{n\hat{T}}^{(n+1)\hat{T}} s_m(n)\mathrm{e}^{\mathrm{j}\frac{2\pi}{\Delta}mt} s^*_{m+l}(n)\mathrm{e}^{-\mathrm{j}\frac{2\pi}{\Delta}(m+l)t}\mathrm{d}t\\ &\qquad = s_m(n)s^*_{m+l}(n)\frac{1}{\hat{T}}\int_{n\hat{T}}^{(n+1)\hat{T}} \mathrm{e}^{\mathrm{j}\frac{2\pi}{\Delta}mt}\mathrm{e}^{-\mathrm{j}\frac{2\pi}{\Delta}(m+l)t}\mathrm{d}t\\ &\qquad = s_m(n)s^*_{m+l}(n)\frac{1}{\hat{T}}\int_{n\hat{T}}^{(n+1)\hat{T}} \mathrm{e}^{-\mathrm{j}\frac{2\pi}{\Delta}lt}\mathrm{d}t\\ &\qquad = \begin{cases} |s_m(n)|^2 & \text{if } l = 0\\ 0 & \text{if } l \neq 0\end{cases},\end{aligned} \tag{3.11}$$

provided $\hat{T}$ is a multiple of Δ, that is, $\hat{T} = \kappa\Delta$, with κ being a positive integer. Indeed, the last equality follows easily by considering that $\mathrm{e}^{-\mathrm{j}\frac{2\pi}{\Delta}lt}$ can be rewritten as

$$\mathrm{e}^{-\mathrm{j}\frac{2\pi}{\Delta}lt} = \cos\left(-\frac{2\pi}{\Delta}lt\right) + \mathrm{j}\sin\left(-\frac{2\pi}{\Delta}lt\right), \tag{3.12}$$

and remembering that both sine and cosine functions integrate to zero in intervals corresponding to multiples of their fundamental period, which in this case is given by $\frac{\Delta}{l}$. Therefore, we must choose $\hat{T}$ in Equation (3.11) in such a way that the cross-correlation is equal to zero for all $l \in \mathcal{M} \setminus \{0\}$. This implies that the choice of $\hat{T}$ must be based on the slowest complex exponential,[3] which occurs when $l = 1$, which in turn shows that $\hat{T}$ must be a multiple of $\frac{\Delta}{1} = \Delta$.

Therefore, as pointed out in Equation (3.11), the orthogonality among subcarriers plays a key role in the choice of $\hat{T}$ and its relation with Δ. Indeed, this orthogonality can be obtained by choosing

[3]The one that takes more time to complete a cycle. In addition, note that the fundamental period of the slowest complex exponential is already a multiple of the periods of the other exponentials.

an OFDM symbol duration $\hat{T} = \kappa\Delta$. Note that, as κ increases, the OFDM symbol duration also increases, but the amount of transmitted data is exactly the same that would be transmitted if $\kappa = 1$, that is, $\kappa > 1$ reduces the system throughput. That is why $\kappa = 1$ is the natural choice. In addition, note that the orthogonality does not depend on the symbols that modulate the subcarriers.

This implies that the OFDM symbol $\hat{u}_n(t)$ can be redefined as

$$\begin{aligned}
\hat{u}_n(t) &\triangleq \frac{1}{\sqrt{\hat{T}}} \sum_{m\in\mathcal{M}} s_m(n) e^{j\frac{2\pi}{\hat{T}} m(t-n\hat{T})}, \quad \text{for } n\hat{T} \leq t < (n+1)\hat{T}, \\
&= \frac{1}{\sqrt{\hat{T}}} \sum_{m\in\mathcal{M}} s_m(n) e^{j\frac{2\pi}{\hat{T}} mt} \underbrace{e^{-j2\pi mn}}_{=1}, \quad \text{for } n\hat{T} \leq t < (n+1)\hat{T}, \\
&= \sum_{m\in\mathcal{M}} s_m(n) \underbrace{\hat{p}(t-n\hat{T}) e^{j\frac{2\pi}{\hat{T}} mt}}_{\triangleq \hat{\varphi}_m(t-n\hat{T})} \\
&= \sum_{m\in\mathcal{M}} s_m(n) \hat{\varphi}_m(t-n\hat{T}),
\end{aligned} \tag{3.13}$$

where function $\hat{\varphi}_m(t)$ represents the mth subcarrier and the pulse signal is

$$\hat{p}(t) \triangleq \begin{cases} \frac{1}{\sqrt{\hat{T}}}, & \text{for } 0 \leq t < \hat{T}, \\ 0, & \text{otherwise.} \end{cases} \tag{3.14}$$

Let us interpret Equation (3.13) pictorially. Each subcarrier $\hat{\varphi}_m(t)$ is a complex exponential multiplied by $\hat{p}(t)$, a rectangular window of duration $\hat{T}$. Figure 3.1(a) depicts a given pulse $\hat{p}(t)$, whose Fourier transform is the well-known sinc function. The square of the sinc represents the subcarrier spectrum, as depicted in Figure 3.1(b). Figure 3.1(c) illustrates many OFDM subcarriers, placed at their correct positions,[4] in order to show the distance between the central frequencies of neighboring subcarriers and to emphasize that, at each subcarrier central frequency, all other subcarriers have amplitude equal to zero, as illustrated by the dotted lines.

Figure 3.1(c) illustrates several subcarriers belonging to a single OFDM symbol as if they were isolated (i.e., we have not added the curves associated with each subcarrier). However, as given in Equation (3.13), an OFDM symbol is formed by the summation of all M subcarriers modulated by their corresponding symbol. The result of such summation is represented in Figure 3.2. This figure depicts a frequency-domain representation of an OFDM symbol comprising three subcarriers. Note that the support of such a representation is the entire real axis.

At this point, it is worth mentioning that simple TDM- and FDM-based transmissions can also yield orthogonal signals to the input of the communication channel, as illustrated in the discussions of Section 1.5 of Chapter 1 within the framework of multiple-access schemes. Indeed,

[4]Remember that the Fourier transform of the product of two signals is equivalent to the convolution between the Fourier transforms of each individual signal. Thus, each subcarrier $\hat{p}(t)e^{j\frac{2\pi}{\hat{T}}mt}$ can be represented in the frequency domain by the convolution of a sinc function (Fourier transform of $\hat{p}(t)$) with a Dirac impulse at frequency $\frac{m}{\hat{T}}$ (Fourier transform of the exponential), which results in the sinc function centered at frequency $\frac{m}{\hat{T}}$.

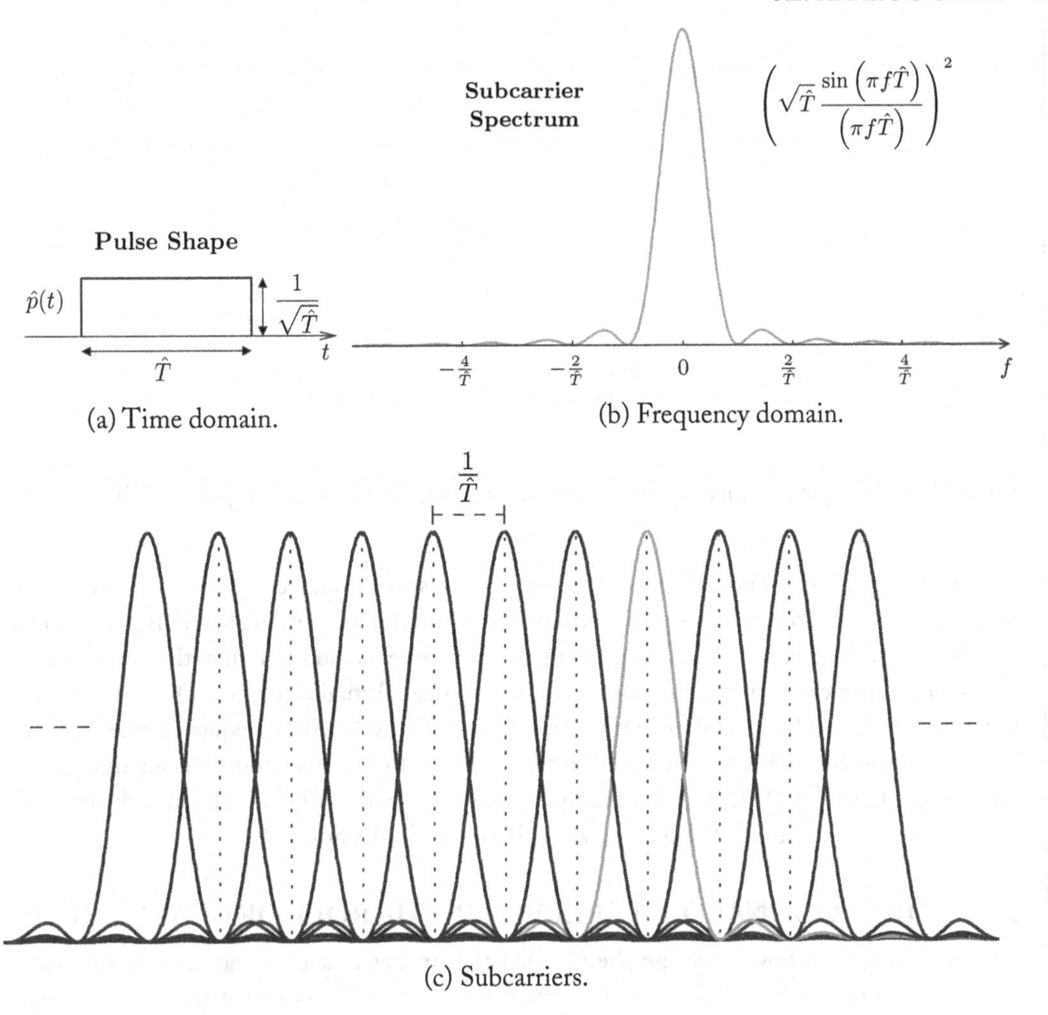

(a) Time domain.

(b) Frequency domain.

(c) Subcarriers.

Figure 3.1: Representation of OFDM subcarriers: (a) time-domain representation of $\hat{p}(t)$; (b) frequency-domain representation of $\hat{p}(t)$; and (c) a set of non-interfering subcarriers.

Equations (3.3) and (3.5) are examples of (theoretically) orthogonal TDM and FDM. However, in the case of FDM transmissions, one must necessarily let empty spectral regions for separating the frequency content associated with each subcarrier, otherwise the filters employed in such separation would be for certain non-causal filters due to the required sharp transitions in the frequency domain. The aforementioned analog OFDM avoids this waste of spectrum that generally occurs in FDM-based systems by allowing spectrum superposition of the subcarriers. In standard TDM transmissions, when the transmitted signal crosses a frequency-selective channel, the original time-domain orthogonality is lost. A possible solution is adding guard intervals between the transmission

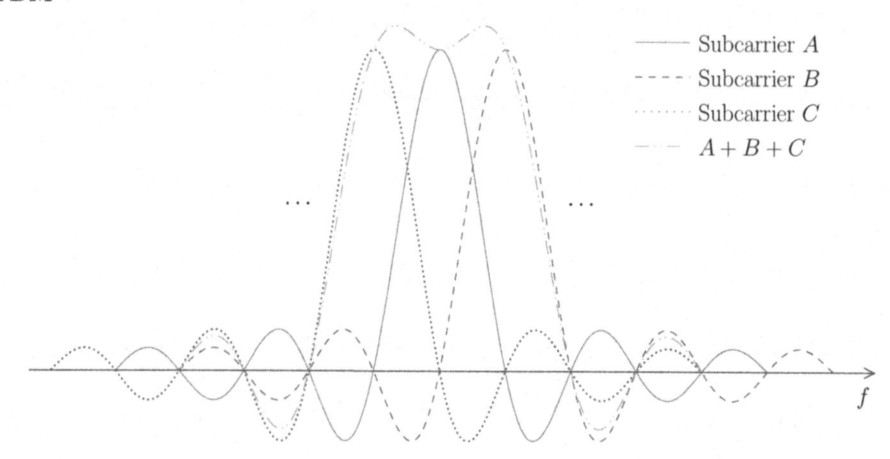

Figure 3.2: Example of frequency representation of an OFDM symbol comprised of three subcarriers.

of each constellation symbol, which also represents a waste of resources. Actually, TDM-based solutions employ time-domain equalizers to decrease the interference among symbols due to the loss of orthogonality. But, even in this case, if the interference level is too high, then the order of the time-domain equalizer can turn its implementation impractical. Actual analog OFDM transmissions, on the other hand, are able to circumvent the interferences introduced by frequency-selective channels by using subcarriers that are orthogonal to each other at the receiver end, thus justifying the name *orthogonal* FDM (OFDM). The key feature present in analog OFDM is the introduction of *guard intervals* between each OFDM symbol, as explained in Subsection 3.2.3.

3.2.3 ORTHOGONALITY AT RECEIVER: THE ROLE OF GUARD INTERVAL

So far we have seen how to design the OFDM transmitter in such a way that its subcarriers are orthogonal to each other, which is an important feature that allows easy extraction of the symbols within an OFDM symbol. However, in practical communications systems this extraction is actually performed at the receiver end. Thus, we are interested in transmitting OFDM symbols that maintain their orthogonality among subcarriers when they reach the receiver end.

The task of keeping the subcarriers orthogonal at the receiver is very challenging, especially when the OFDM symbol faces *multipath fading* channels. As described in Section 1.7, this kind of channel has memory, which means that delayed and attenuated versions of the transmitted signal arrive at the receiver. This generates interference among transmitted constellation symbols and OFDM symbols (assuming the transmission of more than one OFDM symbol). An instinctive solution to maintain the orthogonality and combat interference is to *extend* the time support of the subcarriers.

Let us consider the simplest case in which a subcarrier is a simple sinusoid, as depicted in Figure 3.3, where the solid line represents the original subcarrier with a finite time support and the

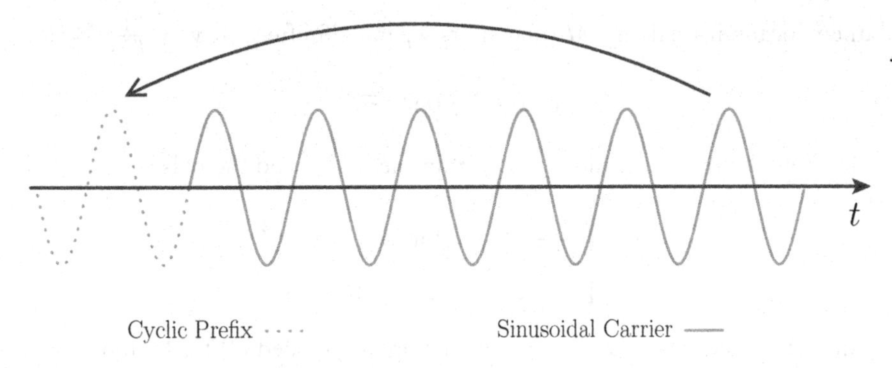

Figure 3.3: Extending an OFDM carrier with a cyclic prefix.

initial dashed version represents its time extension. This OFDM subcarrier extension is known as *cyclic prefix* (CP). Assuming this carrier energizes a frequency-selective channel, whose time-delay spread spans up to $\tau_{\text{mem}} \in \mathbb{R}$ seconds, at the receiver end the first τ_{mem} seconds will be corrupted by the previous OFDM symbol, thus generating the so-called *interblock interference* (IBI). After the period τ_{mem} the IBI is over and if we keep the subcarriers illuminating the channel for T seconds, then the subcarriers will be orthogonal to each other as long as the time period $T - \tau_{\text{mem}}$ is long enough and as long as we are able to eliminate the interference among constellation symbols within a given data block, i.e., the so-called *intersymbol interference* (ISI).

Figure 3.4 illustrates how OFDM symbols are concatenated taking into consideration a guard interval[5] of duration $\tau \in \mathbb{R}$ seconds. The useful symbol time $\hat{T}$ (see Figure 3.4) corresponds to the original duration of the OFDM symbol whose subcarriers are orthogonal to each other only at the transmitter side. Thus, the *extended OFDM symbol* has duration $T \triangleq \hat{T} + \tau$, where the guard interval τ is longer than the longest multipath delay ($\tau > \tau_{\text{mem}}$) and is used to avoid the harmful interferences introduced by frequency-selective channels.

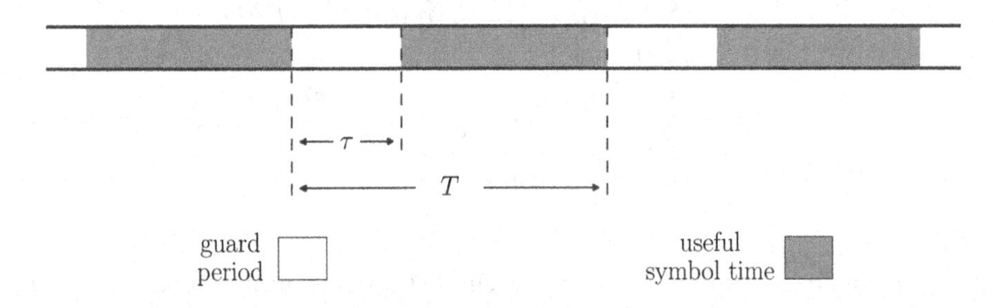

Figure 3.4: Using a guard period to avoid interblock interference.

[5]A guard interval or guard period is a more general concept which includes the cyclic prefix as a special case.

Mathematically, if we define M subcarriers separated in frequency by $\frac{1}{T-\tau}$ Hz as

$$\varphi_m(t) \triangleq p(t)e^{j\frac{2\pi}{T-\tau}mt}, \tag{3.15}$$

for $m \in \mathcal{M}$, where the subcarrier index m is within the set $\mathcal{M}$ and the pulse signal is[6]

$$p(t) \triangleq \begin{cases} \frac{1}{\sqrt{T-\tau}}, & \text{for } -\tau \leq t < T-\tau, \\ 0, & \text{otherwise,} \end{cases} \tag{3.16}$$

then the transmitted signal, which is the concatenation of extended OFDM symbols, can be written as

$$\begin{aligned} u(t) &\triangleq \sum_{n\in\mathbb{Z}} u_n(t) \\ &= \sum_{n\in\mathbb{Z}} \underbrace{\sum_{m\in\mathcal{M}} s_m(n)\varphi_m(t-nT)}_{=u_n(t)}, \end{aligned} \tag{3.17}$$

where $s_m(n)$ is the mth symbol within the nth block representing an extended OFDM symbol.

An extended OFDM symbol obtained by the extension of each subcarrier-time support is equivalent to a time-domain signal with a cyclic prefix of length τ, i.e., in which the first τ seconds of the data block coincide with the last τ seconds. Indeed, consider the nth extended OFDM symbol, whose time support is $[nT-\tau, nT+T-\tau)$. Let t be an arbitrary real number within the first τ seconds of that interval, i.e., $t \in [nT-\tau, nT)$. In addition, let t' be a real number defined as $t' \triangleq t + (T-\tau)$, which denotes a time instant within the last τ seconds of the referred block. Thus, we have

$$\begin{aligned} u_n(t) &= \sum_{m\in\mathcal{M}} s_m(n)p(t-nT)e^{j\frac{2\pi}{T-\tau}m(t-nT)} \\ &= \sum_{m\in\mathcal{M}} s_m(n)\frac{1}{\sqrt{T-\tau}}e^{j\frac{2\pi}{T-\tau}m[t'-(T-\tau)-nT]} \\ &= \sum_{m\in\mathcal{M}} s_m(n)\underbrace{\frac{1}{\sqrt{T-\tau}}}_{=p(t'-nT)}e^{j\frac{2\pi}{T-\tau}m(t'-nT)} \cdot \underbrace{e^{-j2\pi m}}_{=1} \\ &= u_n(t'). \end{aligned} \tag{3.18}$$

A detailed representation of an extended OFDM symbol[7] using cyclic prefix is depicted in Figure 3.5. The reader should note the similarities and differences between Equations (3.13) and (3.17). The signal $\hat{u}_n(t)$ in Equation (3.13) is simply an OFDM symbol, i.e., no guard period is

[6]We considered the interval $[-\tau, T-\tau)$, instead of $[0, T)$, since this choice simplifies the forthcoming notations used in Section 3.3.
[7]In some texts, especially in standards, the extended OFDM symbol is simply referred to as OFDM symbol.

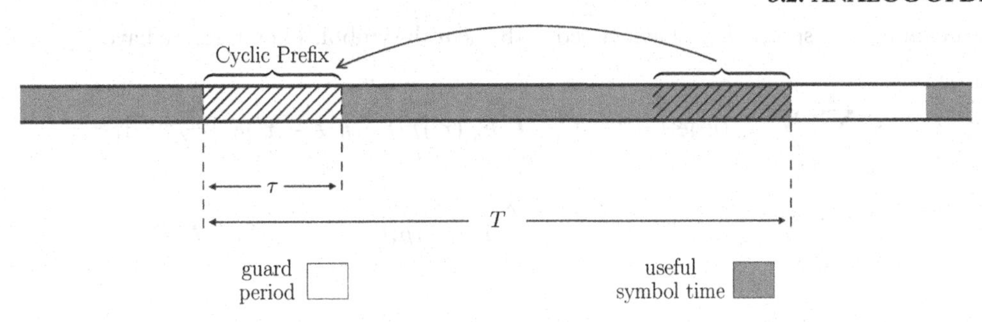

Figure 3.5: Cyclic prefix in an OFDM symbol.

being used and, as a consequence, the received signal corresponding to this $\hat{u}_n(t)$ after crossing a multipath fading channel would not yield orthogonal subcarriers. On the other hand, the signal $u(t)$ in Equation (3.17) is the concatenation of infinitely many extended OFDM symbols using cyclic prefix in order to ensure orthogonality among subcarriers at the receiver side.

Indeed, the orthogonality among subcarriers is maintained since the interference between OFDM symbols (i.e., the IBI) can be eliminated by discarding the first τ seconds out of T seconds of each received data block. As for the remaining interference (ISI) due to constellation-symbol superpositions within the resulting block of duration $T - \tau$, we can eliminate it using the subcarrier orthogonality. In order to verify mathematically these facts, let us analyze the received signal $y(t)$ assuming a noiseless *baseband channel* model whose time-delay spread is τ_{mem}, as follows:

$$\begin{aligned} y(t) &\triangleq \mathcal{H}\{u(t)\} \\ &= \mathcal{H}\left\{\sum_{n'\in\mathbb{Z}}\sum_{m\in\mathcal{M}} s_m(n')\varphi_m(t-n'T)\right\} \\ &= \sum_{n'\in\mathbb{Z}}\sum_{m\in\mathcal{M}} s_m(n')\mathcal{H}\left\{\varphi_m(t-n'T)\right\} \\ &= \sum_{m\in\mathcal{M}}\sum_{n'\in\mathbb{Z}} s_m(n')\mathcal{H}\left\{\varphi_m(t-n'T)\right\}, \end{aligned} \tag{3.19}$$

where $\mathcal{H}\{\cdot\}$ represents the linear system that models the referred baseband channel. Now, by considering that the channel model remains constant during the interval of an OFDM symbol, we can compute the quantity $\mathcal{H}\left\{\varphi_m(t-n'T)\right\}$ through a convolution integral of $\varphi_m(t-n'T)$ with the channel impulse response $h_{n'}(t)$ associated with the n'th symbol. Therefore, we have

$$\begin{aligned} y(t) &= \sum_{m\in\mathcal{M}}\sum_{n'\in\mathbb{Z}} s_m(n')\mathrm{e}^{\mathrm{j}\frac{2\pi}{T-\tau}m(t-n'T)} \int_{-\infty}^{\infty} h_{n'}(\tau')p(t-n'T-\tau')\mathrm{e}^{-\mathrm{j}\frac{2\pi}{T-\tau}m\tau'}\mathrm{d}\tau' \\ &= \sum_{m\in\mathcal{M}}\sum_{n'\in\mathbb{Z}} s_m(n')\mathrm{e}^{\mathrm{j}\frac{2\pi}{T-\tau}m(t-n'T)} \int_{0}^{\tau_{\mathrm{mem}}} h_{n'}(\tau')p(t-n'T-\tau')\mathrm{e}^{-\mathrm{j}\frac{2\pi}{T-\tau}m\tau'}\mathrm{d}\tau', \end{aligned} \tag{3.20}$$

where $h_{n'}(\tau') = 0$ for all $\tau' \notin [0, \tau_{\mathrm{mem}}]$ and[8]

$$p(t-n'T-\tau') = \begin{cases} \dfrac{1}{\sqrt{T-\tau}}, & \text{for } t+\tau-(n'+1)T \le \tau' < t+\tau-n'T, \\ 0, & \text{otherwise.} \end{cases} \tag{3.21}$$

As we are studying both interblock and intersymbol interferences associated with the transmission of blocks $u_n(t)$, with $n \in \mathbb{Z}$, of length T, it is convenient to separate the received signal into blocks with the same length T. Thus, we can write

$$\begin{aligned} y(t) &= \sum_{n\in\mathbb{Z}} \underbrace{y(nT+t')}_{\triangleq y_n(t')} \\ &= \sum_{n\in\mathbb{Z}} y_n(t'), \end{aligned} \tag{3.22}$$

where the time instant t' is in the interval $[-\tau, T-\tau)$. This way, it follows from Equation (3.20) that the nth received block can be expressed by

$$y_n(t') = \sum_{m\in\mathcal{M}}\sum_{n'\in\mathbb{Z}} s_m(n')\mathrm{e}^{\mathrm{j}\frac{2\pi}{T-\tau}m[(n-n')T+t']} I_{n,n'}(t'), \tag{3.23}$$

with

$$I_{n,n'}(t') \triangleq \int_{0}^{\tau_{\mathrm{mem}}} h_{n'}(\tau')p\Big((n-n')T+t'-\tau'\Big)\mathrm{e}^{-\mathrm{j}\frac{2\pi}{T-\tau}m\tau'}\mathrm{d}\tau'. \tag{3.24}$$

Observe that, for $n' \ge n+1$, we have $p((n-n')T+t'-\tau') \ne 0$ only for some values of τ' outside the integration interval $[0, \tau_{\mathrm{mem}}]$. The same occurs for $n' \le n-2$. Indeed, based on Equation (3.21), one has that the pulse signal $p((n-n')T+t'-\tau')$ assumes non-zero values only for values of τ' such that $[n-(n'+1)]T+t'+\tau < \tau' \le (n-n')T+t'+\tau$. Therefore, if $n' \ge n+1$, then the upper bound for τ' will be such that $(n-n')T+t'+\tau \le t'+\tau-T = t'-(T-\tau) < 0$, since $t' \in [-\tau, T-\tau)$, which means that τ' cannot be within the interval $[0, \tau_{\mathrm{mem}}]$ for this

[8] Actually, it would be more appropriate to state that $h_{n'}(\tau')$ is *approximately zero* for all $\tau' \notin [0, \tau_{\mathrm{mem}}]$.

condition. Similarly, if $n' \leq n-2$ then the lower bound for τ' will be such that $[n-(n'+1)]T + t' + \tau \geq t' + \tau + T > \tau > \tau_{\text{mem}}$, which means that τ' cannot be within the interval $[0, \tau_{\text{mem}}]$ in this case as well. Hence, the integrand which appears in the above integral $I_{n,n'}(t')$ is always zero, except for $n' = n$ and $n' = n-1$. When $n' = n$, then the pulse signal $p((n-n')T + t' - \tau')$ is non-zero whenever τ' is such that $t' + \tau - T < 0 < \tau' \leq t' + \tau$. Similarly, when $n' = n-1$ the interval is $0 < t' + \tau < \tau' \leq t' + \tau + T > \tau_{\text{mem}}$. Therefore, we have

$$\begin{aligned} I_{n,n'}(t') =& \frac{\delta[n'-n]}{\sqrt{T-\tau}} \int_{0}^{\min\{t'+\tau,\tau_{\text{mem}}\}} h_{n'}(\tau')\mathrm{e}^{-\mathrm{j}\frac{2\pi}{T-\tau}m\tau'}\mathrm{d}\tau' \\ &+ \frac{\delta[n'-n+1]}{\sqrt{T-\tau}} \int_{\min\{t'+\tau,\tau_{\text{mem}}\}}^{\tau_{\text{mem}}} h_{n'}(\tau')\mathrm{e}^{-\mathrm{j}\frac{2\pi}{T-\tau}m\tau'}\mathrm{d}\tau', \end{aligned} \tag{3.25}$$

where $\delta[n]$ denotes the Kronecker delta, which is defined as

$$\delta[n] \triangleq \begin{cases} 1, & \text{if } n = 0, \\ 0, & \text{otherwise.} \end{cases} \tag{3.26}$$

Now, if we consider only time instants t' within the interval $[0, T-\tau)$, with $\tau > \tau_{\text{mem}}$, then $\min\{t'+\tau, \tau_{\text{mem}}\} = \tau_{\text{mem}}$, and, therefore, the second integral which appears at the right-hand side of Equation (3.25) will be zero. In other words, when we discard the first $\tau > \tau_{\text{mem}}$ seconds of each received block $y_n(t')$, then there is no interference between the nth and $(n-1)$th transmitted OFDM symbols, i.e., for any $t' \in [0, T-\tau)$, we have

$$\begin{aligned} I_{n,n'}(t') &= \frac{\delta[n'-n]}{\sqrt{T-\tau}} \underbrace{\int_{0}^{\tau_{\text{mem}}} h_{n'}(\tau')\mathrm{e}^{-\mathrm{j}\frac{2\pi}{T-\tau}m\tau'}\mathrm{d}\tau'}_{\triangleq H_{n'}\left(\frac{2\pi m}{T-\tau}\right)} \\ &= \frac{\delta[n'-n]}{\sqrt{T-\tau}} H_{n'}\left(\frac{2\pi m}{T-\tau}\right), \end{aligned} \tag{3.27}$$

so that (see Equation (3.23) as well)

$$y_n(t') = \frac{1}{\sqrt{T-\tau}} \sum_{m\in\mathcal{M}} H_n\left(\frac{2\pi m}{T-\tau}\right) s_m(n)\mathrm{e}^{\mathrm{j}\frac{2\pi}{T-\tau}mt'}, \tag{3.28}$$

in which, once again, we highlight that t' must be in the interval $[0, T-\tau)$.

From Equation (3.28), it is clear that, for each fixed time interval $[nT-\tau, nT+T-\tau)$, if one discards the first $\tau > \tau_{\text{mem}}$ seconds of the received signal in that interval, then one ends up with a continuous-time signal that is composed of a sum of M complex exponentials modulated by complex numbers $\frac{s_m(n)H_n\left(\frac{2\pi m}{T-\tau}\right)}{\sqrt{T-\tau}}$. Hence, if we compute the temporal cross-correlation over a useful symbol duration $\hat{T} = T - \tau$ for any of those two modulated subcarriers, we would reach a similar result to Equation (3.11). This means that the orthogonality between subcarriers is achieved at the receiver side. This is the main feature of analog OFDM, since the IBI is eliminated and, at the same time, the ISI can be eliminated by using the resulting orthogonality among subcarriers at the receiver.

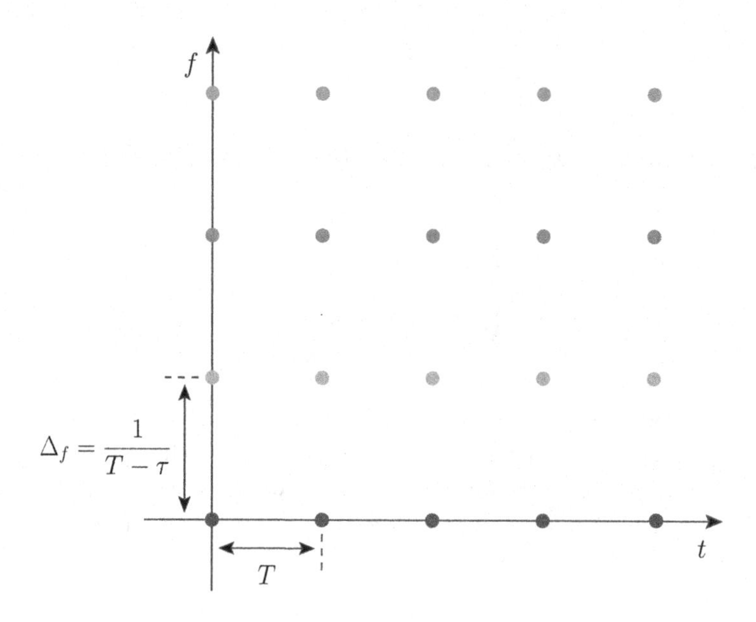

Figure 3.6: Time-frequency map of analog OFDM signals.

Therefore, analog OFDM transmissions illuminate the channel at each $T = \hat{T} + \tau$ period of time, with $\tau > \tau_{\text{mem}}$, using M subcarriers whose central frequencies of neighboring subcarriers are separated by $\Delta_f \triangleq \frac{1}{\hat{T}} = \frac{1}{T-\tau}$, as illustrated in Figure 3.6. Each subcarrier will be responsible for transporting a single symbol during a time slot, taking into consideration the longest path propagation time τ_{mem}.

Note that the extended OFDM symbols do not have orthogonal subcarriers at the transmitter due to the insertion of the guard interval. This can be verified through different ways. For example, if one computes the temporal cross-correlation between two distinct subcarriers described in Equation (3.15), one would end up with a non-zero temporal cross-correlation in general. This occurs because we do not integrate over an integer multiple of the period $T - \tau$. Another way of verifying this fact is to observe Figure 3.7, in which the time-domain and frequency-domain representations of the pulse $p(t)$ in extended OFDM symbols are depicted. As the subcarrier central frequencies are proportional to $\frac{1}{T-\tau}$ and the zeros of the subcarrier spectra are proportional to $\frac{1}{T}$, then there

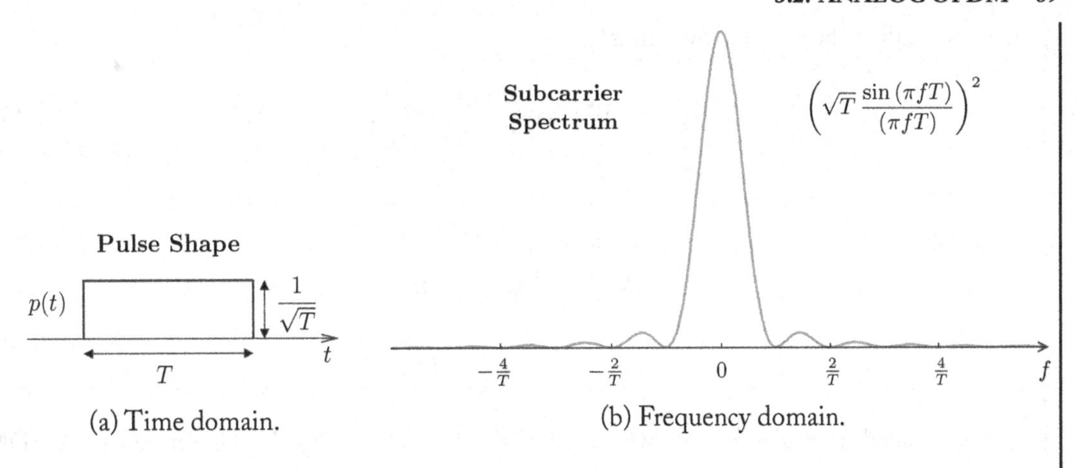

(a) Time domain. (b) Frequency domain.

Figure 3.7: Representation of extended OFDM subcarriers: (a) time-domain representation of $p(t)$; and (b) frequency-domain representation of $p(t)$.

will exist interferences from adjacent subcarriers at those central frequencies. This means that the subcarriers are not orthogonal at the transmitter side. However, this is not an issue since the information extraction occurs at the receiver. Thus, the orthogonality among subcarriers at the receiver end allows a proper extraction of the symbols associated with different subcarriers within a received extended OFDM symbol, even when the received signal has been severely distorted by the channel. That is why we exchange the original orthogonality present in OFDM symbols by an orthogonality of the received extended OFDM symbols with cyclic prefix (after discarding the first τ seconds).

3.2.4 SPECTRAL EFFICIENCY, PAPR, CFO, AND I/Q IMBALANCE

Before we move on to describe an implementation sketch of analog OFDM, let us briefly comment on its *spectral efficiency* and some of the OFDM issues, namely *peak-to-average power ratio* (PAPR), *carrier-frequency offset* (CFO), and *in-phase/quadrature-phase* (I/Q) *imbalance.*

Let us start with the spectral efficiency. We know that, during a period of T seconds, OFDM transmits M symbols from a given constellation with 2^b points, where b is a natural number denoting the number of bits required to represent a single symbol. Thus, the OFDM bit rate (BR) is

$$\mathrm{BR} = \frac{Mb}{T} \tag{3.29}$$

in bits per second (bps). If we add the spectra of all subcarriers and consider that side lobes below 20 dB from the main lobe are negligible, corresponding to the second side lobes on each side of the main lobe, then the total bandwidth is[9]

$$\text{BW} = \frac{M-1}{T-\tau} + 2\frac{3}{T}. \tag{3.30}$$

The ratio

$$\begin{aligned}\frac{\text{BR}}{\text{BW}} &= \frac{Mb}{\frac{(M-1)T}{T-\tau} + 6} \\ &= \frac{b}{\left(\frac{M-1}{M}\right)\frac{T}{T-\tau} + \frac{6}{M}}\end{aligned} \tag{3.31}$$

is the so-called spectral efficiency, which tends to $b(1 - \frac{\tau}{T})$ for large M. That means the OFDM is an optimal modulation in terms of spectral efficiency as long as $\frac{\tau}{T} \ll 1$. However, when $\tau \lessapprox T$ (very dispersive environment), then the spectral efficiency of OFDM transmissions is quite small. Other multicarrier and single-carrier transmissions which address this drawback of spectral efficiency will be described in Chapters 4 and 5.

Figure 3.8 depicts the instantaneous power of a given OFDM symbol $u_n(t)$. A dotted line is used to represent the average power of the OFDM symbol. In this figure, we can observe that there exist some peaks in the power of $u_n(t)$ that are much higher than the average power, i.e., they are well above the dotted line. Indeed, it is well known that the peak-to-average power ratio (PAPR) of OFDM transmissions is higher than the PAPR of single-carrier transmissions (see Subsection 3.4.1 for further details in the discrete-time domain). High PAPR is undesirable because it implies a wide dynamic range of the signal to be transmitted, which in turn requires power amplifiers with linear response over a wide range, increasing the cost of such devices. This is one of the main reasons why in LTE the use of OFDMA in the uplink was avoided. Therefore, PAPR is an important impairment related to OFDM transmissions.

In cases where the carrier frequency of the received signal does not match the carrier frequency of the transmitted signal, we have the so-called carrier-frequency offset (CFO). Thus, CFO is the offset (difference) between two numbers representing carrier frequencies, one at the transmitter and the other at the receiver end. Ideally, CFO should be close to zero, but there are many practical cases in which non-negligible CFO occurs. For example, when the transmitter and/or receiver are moving, which usually happens in mobile communications, the Doppler effect acts as a source of CFO. Note that, from our previous discussion about the importance of orthogonality in OFDM systems, it is rather intuitive that CFO has the potential to severely degrade the quality of OFDM transmissions. Most of the solutions to the CFO issue rely on blind estimation of the frequency offset, and are a bit complex and/or applicable to very particular cases. Some of the low-complexity solutions to CFO are presented in [46, 92].

[9]The bandwidth must be computed by considering that the central frequencies of the M subcarriers are separated by $\frac{1}{T-\tau}$ Hz, and each subcarrier is a sinc whose second side lobes decay more than 20 dB at $\frac{3}{T}$ Hz, as can be observed in Figure 3.7(b).

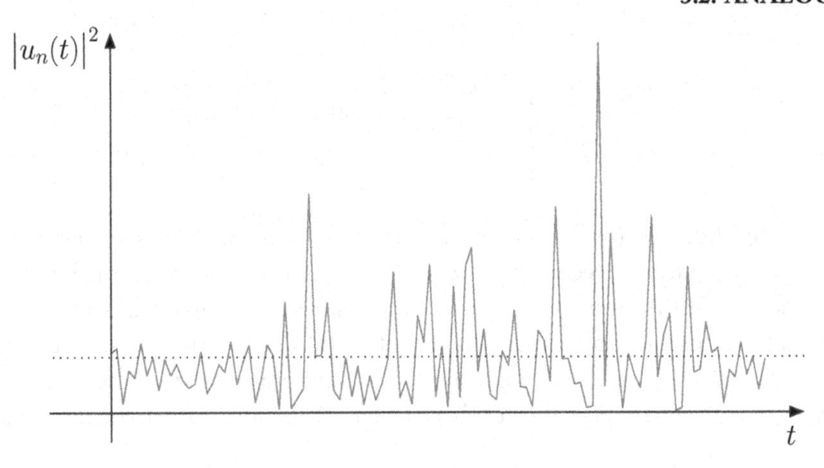

Figure 3.8: Instantaneous power of a single OFDM symbol.

Moreover, digital transmissions usually employ two branches: an in-phase (I) and a quadrature-phase (Q) branch. These branches are associated with the real and imaginary parts of the transmitted signal, respectively. I/Q imbalance occurs when there is phase and/or amplitude mismatches between I and Q branches. Such mismatches are usually due to the imperfections in the process of the radio-frequency signal down-conversion to baseband signal and are, therefore, unavoidable in the analog front-end [74]. In most cases, I/Q imbalance can only deteriorate the bit-error rate (BER) performance of OFDM systems when they are employing high-order modulation schemes, such as 64-QAM (quadrature amplitude modulation). When I/Q imbalance is a major issue, one can use digital signal processing techniques to compensate such mismatch. Indeed, there already exists several techniques to compensate for the I/Q imbalance without increasing significantly the computational burden. For more details, see [6, 78, 85] and references therein.

3.2.5 IMPLEMENTATION SKETCH

Figure 3.9 depicts the implementation sketch of an analog OFDM transmission scheme. First the symbols $s_m(n)$ modulate their corresponding time-extended subcarriers $\varphi_m(t - nT)$. This operation is represented by the filtering of the continuous-time signal $s_m(n)\delta(t - nT)$ through the analog filter $\varphi_m(t)$. The signals resulting from each of these modulations are then added, forming an extended OFDM symbol, which crosses an analog channel and is then corrupted by the environment noise $v(t)$. Here, we assume that the length of the cyclic prefix is greater than the channel time-delay spread. At the receiver end, there are filters $\psi_m(t)$ which are responsible for discarding the cyclic prefix as well as extracting the symbol $s_m(n)$ from the corrupted OFDM symbol. Indeed, by defining the receiver filters as

$$\psi_m(t) \triangleq q(t)e^{\frac{j2\pi}{T-\tau}mt}, \tag{3.32}$$

for $m \in \mathcal{M}$, where

$$q(t) \triangleq \begin{cases} \frac{1}{\sqrt{T-\tau}}, & \text{for } 0 \leq t < T-\tau, \\ 0, & \text{otherwise,} \end{cases} \tag{3.33}$$

the output of the filter $\psi_m(t)$ will be equivalent to the temporal cross-correlation between the OFDM symbol and the mth receiver filter. Following the same steps performed in Equation (3.11) and remembering that the fundamental period of the slowest subcarrier is $\Delta = T - \tau$, it is easy to verify that $\psi_m(t)$ will remove all subcarriers except $\varphi_m(t)$ due to orthogonality. In fact, the pair of functions $\varphi_m(t)$ and $\psi_m(t)$ are biorthogonal [17]. In addition, note that the basis function at the receiver $\psi_m(t)$ has a time support shorter than the basis function at the transmitter $\varphi_m(t)$.

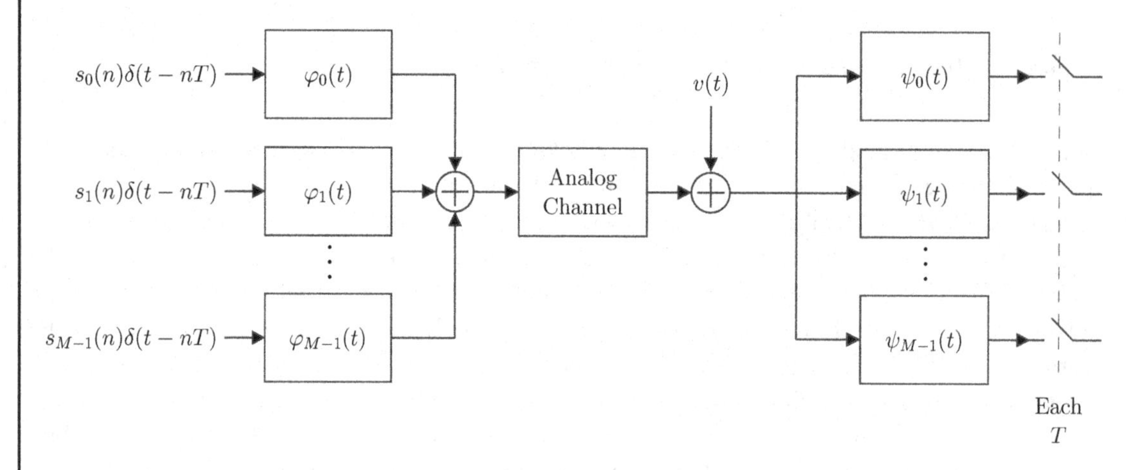

Figure 3.9: Analog OFDM implementation sketch.

Even though analog OFDM can be derived from a very insightful view of the digital-to-analog conversion implemented in an orthogonal FDM-based fashion, the resulting implementation sketch depicted in Figure 3.9 also summarizes its main drawback: in general, practical solutions entail the use of a large number of orthogonal subcarriers, thus hindering the applicability of this structure in practice. Indeed, if M is large, then we would have to implement a large amount of different oscillators/modulators, which may not be practical. This is one of the main reasons why this analog version of OFDM was not employed in commercial applications after its proposal by R. W. Chang [10, 11]. However, many of its properties and interpretations are still useful, as S. B. Weinstein and P. M. Ebert [89], as well as A. Peled and A. Ruiz [64] noticed when they realized that OFDM could be efficiently implemented in the discrete-time domain. This implementation will be addressed in the next section.

3.3 DISCRETE-TIME OFDM

Digital signal processing (DSP) has emerged as a powerful and efficient tool in a growing number of applications. Indeed, there are many situations where the use of DSP-based techniques has either greatly simplified the implementation of practical systems or simply enhanced their performance. The discrete-time implementation of OFDM systems is an example of how DSP can even enable the practical usage of a given technique, which could be quite hard to implement otherwise.

In this section we will apply sampling to the OFDM symbols and show a discrete-time implementation of OFDM systems. As will be shown, the discrete Fourier transform (DFT) plays a central role in this process, enabling efficient implementations of OFDM by means of fast Fourier transform (FFT) algorithms.

3.3.1 DISCRETIZATION OF THE OFDM SYMBOL

As in the continuous-time case, let us start with OFDM symbols without guard intervals described in Subsection 3.2.2. The first challenge that one faces while trying to derive a discrete-time implementation of OFDM is that the *OFDM symbols are not bandlimited.* Indeed, due to the use of a time-domain rectangular window $\hat{p}(t)$ as illustrated in Figure 3.1(a), each subcarrier in an OFDM symbol and the entire OFDM symbol itself contain spectral components at infinitely large frequencies, as depicted in Figures 3.1(b) and 3.1(c). Therefore, a uniform sampling of such continuous-time signal would invariably result in *aliasing effects*, regardless of the particular sampling frequency employed.

In other words, no discrete-time implementation of OFDM obtained by sampling the analog OFDM can be used to recover the same original continuous-time signal associated with analog OFDM (which employs rectangular windows). In general, the recovering process is implemented as a digital-to-analog conversion, such as the one described in expression (3.1), with the constellation symbols being replaced by the elements that compose an OFDM symbol. Such impossibility is also pointed out in [43].

In principle, the presence of aliasing effects could hinder any interpretation based on the intuitive results of analog OFDM presented in Section 3.2. Fortunately, the aliasing effects are not determinant here and the reason is quite simple to understand: the original signal to be transmitted, $s_m(n)$, has a discrete nature, which means that we are not interested in the particular form of the continuous-time OFDM symbol as long as we can detect the original discrete-time signal.[10] It turns out that each element of this discrete signal is the scaling factor of the subcarrier spectra at their central frequencies. Hence, if the spectral components at the subcarrier central frequencies do not suffer from aliasing effects, then we could recover the discrete-time signal $s_m(n)$. Once again, the orthogonality of the subcarriers in an OFDM symbol plays an important role here (see Figure 3.1(c)).

[10] This is true only for detection purposes. If one is interested in studying the spectral roll-off related to the output of OFDM systems or the effect of CFO (just to mention a few examples), then one should work directly with the discrete-time OFDM model after the digital-to-analog conversion [43].

Indeed, when we sample an OFDM symbol,[11] the spectral repetitions due to the sampling process are spaced apart by f_s Hz, where $f_s \triangleq \frac{1}{T_s}$ denotes the sampling frequency and $T_s \in \mathbb{R}$ denotes the sampling period. First of all, f_s must be larger than $\frac{M-1}{\hat{T}}$, otherwise we would have at least one central frequency being shared by two distinct subcarriers, constituting a harmful type of aliasing in this case. But if we consider any integer multiple of $\frac{1}{\hat{T}}$ larger than $M-1$, then we would not have any kind of interference at the subcarrier central frequencies. For simplicity reasons, it is better to use the smallest sampling frequency that does not cause interference at the subcarrier central frequencies, i.e., $f_s = \frac{M}{\hat{T}}$. Thus, we have

$$T_s \triangleq \frac{\hat{T}}{M}. \tag{3.34}$$

Without loss of generality, since the OFDM symbols are non-overlapping in time, we can analyze each block of symbols separately. The nth OFDM symbol is non-zero only for time instants within the interval $[n\hat{T}, (n+1)\hat{T})$. This means that, for all $t' \in [0, \hat{T})$, we have from Equation (3.13) that

$$\begin{aligned}\hat{u}_n(t'+n\hat{T}) &= \sum_{m\in\mathcal{M}} s_m(n)\hat{\varphi}_m(t'+n\hat{T}-n\hat{T})\\ &= \sum_{m\in\mathcal{M}} s_m(n)\hat{\varphi}_m(t').\end{aligned} \tag{3.35}$$

Now, let $\hat{u}_{n,k}$ be the discrete-time signal stemming from sampling the continuous-time signal $\hat{u}_n(t'+n\hat{T})$ described in the above equation at each time instant $t' = kT_s$, where, due to the definition of T_s in Equation (3.34) and as $t' \in [0, \hat{T})$, we must have k within the set $\mathcal{M}$. Hence, the resulting discrete-time representation of an OFDM symbol is

$$\begin{aligned}\hat{u}_{n,k} &\triangleq \hat{u}_n(kT_s+n\hat{T})\\ &= \sum_{m\in\mathcal{M}} s_m(n)\underbrace{\hat{\varphi}_m(kT_s)}_{\triangleq\hat{\varphi}_{m,k}}\\ &= \sum_{m\in\mathcal{M}} s_m(n)\hat{\varphi}_{m,k},\end{aligned} \tag{3.36}$$

where, based on Equation (3.13), we have that the discrete-time version of the mth subcarrier in an OFDM symbol is given as

$$\begin{aligned}\hat{\varphi}_{m,k} &= \frac{1}{\sqrt{\hat{T}}}\mathrm{e}^{\mathrm{j}\frac{2\pi}{\hat{T}}m\left(\frac{k\hat{T}}{M}\right)}\\ &= \frac{1}{\sqrt{\hat{T}}}\mathrm{e}^{\mathrm{j}\frac{2\pi}{M}mk}.\end{aligned} \tag{3.37}$$

[11] Here, sampling means to multiply the continuous-time function by a train of Dirac impulses.

Thus, we can rewrite the discrete OFDM symbol in a more convenient manner as follows:

$$\begin{aligned}\hat{u}_{n,k} &= \frac{1}{\sqrt{\hat{T}}} \underbrace{\sum_{m \in \mathcal{M}} s_m(n) e^{j\frac{2\pi}{M}mk}}_{\triangleq \sqrt{M}\mathrm{IDFT}\{s_m(n)\}_k} \\ &= \sqrt{\frac{M}{\hat{T}}} \mathrm{IDFT}\left\{s_m(n)\right\}_k,\end{aligned} \tag{3.38}$$

for each $k \in \mathcal{M}$, where the inverse DFT (IDFT) of the discrete-time signal $s_m(n)$, with $m \in \mathcal{M}$, is also a sequence with length M. Hence, the following relation also holds:

$$s_m(n) = \sqrt{\frac{\hat{T}}{M}} \mathrm{DFT}\left\{\hat{u}_{n,k}\right\}_m, \tag{3.39}$$

in which the DFT of the sequence $\hat{u}_{n,k}$ is also a sequence whose mth element is defined as

$$\mathrm{DFT}\left\{\hat{u}_{n,k}\right\}_m \triangleq \frac{1}{\sqrt{M}} \sum_{k \in \mathcal{M}} \hat{u}_{n,k} e^{-j\frac{2\pi}{M}mk}, \tag{3.40}$$

thus implying that the symbols $s_m(n)$ can be actually written as

$$s_m(n) = \frac{\hat{T}}{M} \sum_{k \in \mathcal{M}} \hat{u}_{n,k} \hat{\varphi}^*_{m,k}. \tag{3.41}$$

In summary, the discrete-time version of OFDM symbols without guard intervals is easily computed through an IDFT of the transmitted discrete-time signal, which can be efficiently implemented by using an FFT algorithm. Thus, assuming the channel introduces no distortion on the transmitted signal, the sequence of symbols $s_m(n)$ could be recovered at the receiver end by taking the DFT of the sequence $\hat{u}_{n,k}$ and scaling the result, as shown in Equation (3.39). This is a rather obvious conclusion since the DFT and IDFT are inverse operations.

Note that the discrete-time OFDM symbols $\hat{u}_{n,k}$ in expression (3.36) can be thought as the superposition of M subcarriers $\hat{\varphi}_{m,k}$ modulated by the symbols $s_m(n)$. It is worth mentioning that the original orthogonality present in the analog OFDM symbol is preserved in the discrete-time case. Indeed, we can see that the following relation between any two subcarriers with indexes $i, j \in \mathcal{M}$

is valid:

$$
\begin{aligned}
\sum_{k\in\mathcal{M}} \hat{\varphi}_{j,k}\hat{\varphi}^*_{i,k} &= \frac{1}{\hat{T}} \sum_{k\in\mathcal{M}} e^{j\frac{2\pi}{M}k(j-i)} \\
&= \begin{cases} \frac{1}{\hat{T}}\left(\frac{e^{j\frac{2\pi}{M}M(j-i)}-1}{e^{j\frac{2\pi}{M}(j-i)}-1}\right), & \text{if } i \neq j, \\ \frac{M}{\hat{T}}, & \text{otherwise.} \end{cases} \\
&= \begin{cases} 0, & \text{if } i \neq j, \\ \frac{M}{\hat{T}}, & \text{otherwise.} \end{cases} \\
&= \frac{M}{\hat{T}}\delta[i-j].
\end{aligned} \tag{3.42}
$$

Equation (3.42) is nothing but the orthogonality between subcarriers which are synchronized and have the same duration $\hat{T}$. This expression stems from the projection of a transmitted signal onto a subcarrier for detection purposes at the receiver end, as exemplified in Equation (3.41). For analog OFDM, we have shown that the subcarrier orthogonality at the transmitter side is not sufficient to allow the detection of the transmitted symbols in practical situations, i.e., when the data faces a frequency-selective channel. This occurs since frequency-selective fading channels extend the time support of the transmitted signals generating both IBI and ISI, as explained in Subsection 1.7. In Subsection 3.2.3, we ensure subcarrier orthogonality at the receiver by inserting a cyclic prefix before transmission. In the following subsection, we shall generate a discrete-time version of the results related to extended analog OFDM symbols using cyclic prefix.

3.3.2 DISCRETIZATION AT RECEIVER: THE CP-OFDM

Our starting point in this subsection is the expression (3.28) of the received block after removing the first τ seconds. As pointed out before, τ is chosen in such a manner that it is larger than the delay spread of the channel τ_{mem}. Let L be a positive integer number defined as

$$
L \triangleq \left\lceil \frac{\tau_{\text{mem}}}{T_s} \right\rceil, \tag{3.43}
$$

where $T_s = \frac{\hat{T}}{M}$ is the sampling rate associated with the discretization process. In addition, let K be a positive integer number such that $K \geq L$. Thus, by choosing the length of the cyclic prefix as

$$
\tau \triangleq KT_s, \tag{3.44}
$$

then we have that $\tau > \tau_{\text{mem}}$ and, therefore, expression (3.28) holds for any $t' \in [0, T-\tau)$.

Now, by remembering from Subsection 3.2.3 that the useful symbol time $\hat{T}$ is given by $T-\tau$, then the sampling rate can be written as

$$
T_s = \frac{T-\tau}{M}, \tag{3.45}
$$

which means that $[0, T - \tau) = [0, MT_s)$.

Let $y_{n,k}$ be the discrete-time signal originating from sampling the continuous-time signal $y_n(t')$ described in Equation (3.28) at each time instant $t' = kT_s$, with $k \in \mathcal{M}$. As explained before, this continuous-time signal is the nth received block after removing the first $\tau = KT_s$ seconds. Thus, the resulting discrete-time representation of the nth received extended OFDM symbol after removing the first K elements[12] is

$$\begin{aligned} y_{n,k} &\triangleq y_n(kT_s) \\ &= \frac{1}{\sqrt{T-\tau}} \sum_{m\in\mathcal{M}} H_n\left(\frac{2\pi m}{T-\tau}\right) s_m(n) \mathrm{e}^{\mathrm{j}\frac{2\pi}{T-\tau}mk\frac{T-\tau}{M}} \\ &= \frac{1}{\sqrt{T-\tau}} \sum_{m\in\mathcal{M}} H_n\left(\frac{2\pi m}{T-\tau}\right) s_m(n) \mathrm{e}^{\mathrm{j}\frac{2\pi}{M}mk} \\ &= \sum_{m\in\mathcal{M}} \underbrace{H_n\left(\frac{2\pi m}{T-\tau}\right)}_{\triangleq \lambda_m(n)} s_m(n) \underbrace{\frac{\mathrm{e}^{\mathrm{j}\frac{2\pi}{M}mk}}{\sqrt{T-\tau}}}_{=\hat{\varphi}_{m,k}} \\ &= \sum_{m\in\mathcal{M}} \lambda_m(n) s_m(n) \hat{\varphi}_{m,k}. \end{aligned} \tag{3.46}$$

Equation (3.46) means that the received signal, after removing the first K elements associated with the guard interval introduced at the transmitter side, is composed of M modulated subcarriers $\hat{\varphi}_{m,k}$. Even though the kth element of such signal is affected by all transmitted symbols $s_m(n)$, with $m \in \mathcal{M}$, one can use the subcarrier orthogonality expressed in Equation (3.42) in order to recover a scaled version of the transmitted symbols without ISI. Therefore, by projecting the received signal onto the mth subcarrier, one gets

$$\lambda_m(n) s_m(n) = \frac{T-\tau}{M} \sum_{k\in\mathcal{M}} y_{n,k} \hat{\varphi}^*_{m,k}. \tag{3.47}$$

The former projection process can be implemented in a much more efficient way. Indeed, based on Equation (3.46) and following a similar approach which was employed in Subsection 3.3.1, we can rewrite the received discrete OFDM symbol in a more convenient manner as follows:

$$y_{n,k} = \sqrt{\frac{M}{T-\tau}} \mathrm{IDFT}\left\{\lambda_m(n) s_m(n)\right\}_k, \tag{3.48}$$

for each $k \in \mathcal{M}$. Therefore, the following relation also holds:

$$\lambda_m(n) s_m(n) = \sqrt{\frac{T-\tau}{M}} \mathrm{DFT}\left\{y_{n,k}\right\}_m. \tag{3.49}$$

[12] If we are ignoring the first $\tau = KT_s$ seconds of the continuous-time received block, then we are ignoring the first K elements of the related discrete-time signal with corresponding sampling rate of $\frac{1}{T_s}$ Hz.

Hence, in order to recover $s_m(n)$, all we need is to multiply the mth element of the DFT of the received OFDM symbol by $\frac{1}{\lambda_m(n)}$, assuming $\lambda_m(n) \neq 0$, for all $m \in \mathcal{M}$. This is the so-called *zero-forcing* (ZF) equalizer. In fact, there are many other ways to perform equalization in order to estimate $s_m(n)$. The choice of the equalizer depends on the types of distortion faced by the transmitted signal. For instance, in our previous discussion we have neglected the existence of additive noise. In the presence of such type of noise, an equalizer that minimizes the mean square error (MSE) would be more appropriate than an equalizer that eliminates only the ISI, such as the ZF equalizer. In this case, the equalization would consist of multiplying the mth element of the DFT of the received OFDM symbol by

$$\frac{\lambda_m^*(n)}{|\lambda_m(n)|^2 + \frac{\sigma_v^2(n)}{\sigma_s^2(n)}}, \tag{3.50}$$

where $\sigma_s^2(n)$ and $\sigma_v^2(n)$ represent the variance of symbols and noise, respectively.[13] This type of equalizer is known as minimum MSE (MMSE) equalizer, as previously discussed in Subsection 2.4.1.

As explained in Subsection 3.2.3, the subcarrier orthogonality at the receiver end is obtained by including a cyclic prefix of length τ at the transmitter end, generating the extended OFDM symbols. Thus, by following the same steps employed in Subsection 3.3.1, but now considering the nth extended OFDM symbol $u_n(t)$ of Equation (3.17), we have

$$\begin{aligned} u_n(t' + nT) &= \sum_{m\in\mathcal{M}} s_m(n)\varphi_m(t' + nT - nT) \\ &= \sum_{m\in\mathcal{M}} s_m(n)\varphi_m(t'), \end{aligned} \tag{3.51}$$

for all $t' \in [-\tau, T - \tau) = [-KT_s, MT_s)$. Hence, for each $k \in \{-K, \cdots, -1, 0, \cdots, M-1\}$, we can define the discrete-time representation of an extended OFDM symbol as (see Equation (3.15))

$$\begin{aligned} u_{n,k} &\triangleq u_n(kT_s + nT) \\ &= \frac{1}{\sqrt{T - \tau}} \sum_{m\in\mathcal{M}} s_m(n) e^{j\frac{2\pi}{M}mk}, \end{aligned} \tag{3.52}$$

where, for $k \in \mathcal{M}$, one has

$$\sum_{m\in\mathcal{M}} s_m(n) e^{j\frac{2\pi}{M}mk} = \sqrt{M}\text{IDFT}\{s_m(n)\}_k, \tag{3.53}$$

[13] It is assumed here that all constellation symbols within the nth OFDM symbol have the same variance $\sigma_s^2(n)$. In addition, it was assumed that all noise components have the same variance $\sigma_v^2(n)$ as well. Those assumptions are not necessary, but they simplify the notation. See Subsection 3.4.4 for the case where we do not consider those equal-power assumptions.

whereas, for $k \in \{-K, \cdots, -1\}$, one has

$$\begin{aligned}\sum_{m\in\mathcal{M}} s_m(n)e^{j\frac{2\pi}{M}mk} &= \sum_{m\in\mathcal{M}} s_m(n)e^{j\frac{2\pi}{M}mk} \underbrace{e^{j\frac{2\pi}{M}mM}}_{=1} \\ &= \sum_{m\in\mathcal{M}} s_m(n)e^{j\frac{2\pi}{M}(M+k)m} \\ &= \sqrt{M}\text{IDFT}\{s_m(n)\}_{(M+k)},\end{aligned} \tag{3.54}$$

with $M + k \in \{M - K, \cdots, M - 1\} \subset \mathcal{M}$, assuming that $M \geq K$, i.e., the useful symbol duration is larger than or equal to the duration of the guard interval.

Therefore, we can use these results in Equation (3.52) so that, for each $k \in \{-K, \cdots, -1, 0, \cdots, M - 1\}$, we have

$$u_{n,k} = \sqrt{\frac{M}{T-\tau}}\text{IDFT}\{s_m(n)\}_k, \tag{3.55}$$

in which the above notation considers the inherent periodicity property of the IDFT, i.e., $\text{IDFT}\{s_m(n)\}_{(-k)} = \text{IDFT}\{s_m(n)\}_{(M-k)}$, for $k \in \{1, \cdots, K\}$. Therefore, the first K elements of the signal $u_{n,k}$ are equal to its last K elements, thus characterizing the type of guard interval as *cyclic prefix* (CP-OFDM). In the discrete-time domain, this guard interval is also called *redundancy* since it corresponds to entries that do not carry additional information.

So far, we described the discrete-time versions of the transmitter and receiver. We shall address the channel model in Subsection 3.3.3.

3.3.3 DISCRETE-TIME MULTIPATH CHANNEL

The term $\lambda_m(n)$ which appears in expression (3.46) is associated with the frequency response of the analog channel evaluated at the central frequency $\frac{m}{T-\tau}$ Hz, for each subcarrier index $m \in \mathcal{M}$ and for each block $n \in \mathbb{Z}$. The aim of this subsection is to show the relation between $\lambda_m(n)$ and the discrete-time model of the analog baseband channel.

In actual discrete-time OFDM implementations, one must always associate the discrete OFDM signal $u_{n,k}$ with a related continuous-time signal, let us say $\overline{u}(t)$. As already mentioned, due to aliasing effects this continuous-time version is not the analog OFDM symbol presented in Section 3.2. The standard way to perform such conversion is through a DAC process, as briefly described in Subsection 3.2.1, so that

$$\overline{u}(t) \triangleq \sum_{n\in\mathbb{Z}} \sum_{k=-K}^{M-1} u_{n,k} p_{\text{T}}(t - (k + nN)T_s), \tag{3.56}$$

where

$$N \triangleq M + K \tag{3.57}$$

denotes the amount of transmitted elements per extended OFDM symbol, T_s is the sampling period of the DAC, and $p_{\rm T}(t)$ is the *transmitting pulse* that may be chosen based on several distinct criteria, such as best spectral roll-off.

After applying $\overline{u}(t)$ to a continuous-time baseband channel represented by the linear operator $\overline{\mathcal{H}}\{\cdot\}$, we have the following resulting signal:

$$\begin{aligned} r(t) &\triangleq \overline{\mathcal{H}}\{\overline{u}(t)\} \\ &= \sum_{n\in\mathbb{Z}} \sum_{k=-K}^{M-1} u_{n,k} \overline{\mathcal{H}}\{p_{\rm T}(t-(k+nN)T_s)\}, \end{aligned} \tag{3.58}$$

where $\overline{\mathcal{H}}\{p_{\rm T}(t-(k+nN)T_s)\}$ is the output of the linear channel corresponding to nth transmitted block, which can be computed as

$$\overline{\mathcal{H}}\{p_{\rm T}(t-(k+nN)T_s)\} = (\overline{h}_n * p_{\rm T})(t-(k+nN)T_s). \tag{3.59}$$

The function $\overline{h}_n(t)$ represents the impulse response of the analog baseband channel associated with the nth transmitted block.[14] Hence, one has

$$r(t) = \sum_{n\in\mathbb{Z}} \sum_{k=-K}^{M-1} u_{n,k} (\overline{h}_n * p_{\rm T})(t-(k+nN)T_s). \tag{3.60}$$

At the receiver end, an analog-to-digital conversion is also implemented by first filtering the received waveform and then sampling it at a given rate (e.g., the same sampling rate employed at the transmitter). By considering that the *receiving filter* impulse response is $p_{\rm R}(t)$, then the signal $\overline{y}(t)$ resulting from this first filtering stage at the receiver front-end is given by

$$\begin{aligned} \overline{y}(t) &\triangleq (p_{\rm R} * r)(t) \\ &= \sum_{n\in\mathbb{Z}} \sum_{k=-K}^{M-1} u_{n,k} \underbrace{(p_{\rm R} * \overline{h}_n * p_{\rm T})(t-(k+nN)T_s)}_{\triangleq h_n(t-(k+nN)T_s)} \\ &= \sum_{n'\in\mathbb{Z}} \sum_{k=-K}^{M-1} u_{n,k} h_n(t-(k+nN)T_s), \end{aligned} \tag{3.61}$$

where the *equivalent analog baseband channel model* $h_n(t)$ is given by

$$h_n(t) = (p_{\rm R} * \overline{h}_n * p_{\rm T})(t). \tag{3.62}$$

[14]It is assumed that the channel model does not vary with k during the transmission of the nth block.

We can now define the discrete-time version of $\overline{y}(t)$ as

$$
\begin{aligned}
y_{n',k'} &\triangleq \overline{y}(k'T_s + n'NT_s) \\
&= \sum_{n\in\mathbb{Z}} \sum_{k=-K}^{M-1} u_{n,k} \underbrace{h_n\left(\left(k'-k\right)T_s + (n'-n)NT_s\right)}_{\triangleq h_n^{\mathrm{d}}((k'-k)+(n'-n)N)} \\
&= \sum_{n\in\mathbb{Z}} \sum_{k=-K}^{M-1} u_{n,k} h_n^{\mathrm{d}}\left((k'-k)+(n'-n)N\right),
\end{aligned}
\tag{3.63}
$$

in which the impulse response

$$
\begin{aligned}
h_n^{\mathrm{d}}(l) &\triangleq h_n(lT_s) \\
&= (p_{\mathrm{R}} * \overline{h}_n * p_{\mathrm{T}})(lT_s)
\end{aligned}
\tag{3.64}
$$

is the *discrete-time equivalent channel model* associated with the nth transmitted block.

Now, in order to determine the relation between $\lambda_m(n)$ and $h_n^{\mathrm{d}}(l)$, consider that $h_n(t)$ is the impulse response of a multipath fading channel with delay spread $\tau_{\mathrm{mem}} \leq LT_s$. This way, it can be approximated as[15]

$$
h_n(t) = \sum_{l\in\mathcal{L}} h_n^{\mathrm{d}}(l)\delta(t - lT_s),
\tag{3.65}
$$

in which the set of channel-tap indexes $\mathcal{L}$ is defined as

$$
\mathcal{L} \triangleq \{0, 1, \cdots, L\}.
\tag{3.66}
$$

[15] We use the word "approximated," since the delay associated with the lth tap is not necessarily equal to lT_s. However, if T_s is small enough, then this expression is a reasonable approximation for an equivalent multipath baseband channel whose corresponding transmitting and receiving filters were properly designed.

Taking this fact into account, assuming that $T_s = \frac{T-\tau}{M}$, and considering Equations (3.27) and (3.46), we have

$$\begin{aligned}
\lambda_m(n) &= H_n\left(\frac{2\pi}{T-\tau}m\right) \\
&= \int_0^{\tau_{\text{mem}}} h_n(\tau')\mathrm{e}^{-\mathrm{j}\frac{2\pi}{T-\tau}m\tau'}\mathrm{d}\tau' \\
&= \sum_{l\in\mathcal{L}} h_n^{\mathrm{d}}(l)\int_0^{\tau_{\text{mem}}} \delta(\tau'-lT_s)\mathrm{e}^{-\mathrm{j}\frac{2\pi}{T-\tau}m\tau'}\mathrm{d}\tau' \\
&= \sum_{l\in\mathcal{L}} h_n^{\mathrm{d}}(l)\mathrm{e}^{-\mathrm{j}\frac{2\pi}{T-\tau}mlT_s} \\
&= \sum_{l\in\mathcal{L}} h_n^{\mathrm{d}}(l)\mathrm{e}^{-\mathrm{j}\frac{2\pi}{M}ml} + 0 \\
&= \sum_{l\in\mathcal{L}} h_n^{\mathrm{d}}(l)\mathrm{e}^{-\mathrm{j}\frac{2\pi}{M}ml} + \sum_{k\in\mathcal{M}\backslash\mathcal{L}} 0\times\mathrm{e}^{-\mathrm{j}\frac{2\pi}{M}mk} \\
&= \sqrt{M}H_n^{\mathrm{d}}\left(\mathrm{e}^{\mathrm{j}\frac{2\pi}{M}m}\right) \\
&\triangleq \sqrt{M}H_n^{\mathrm{d}}(m),
\end{aligned} \tag{3.67}$$

in which $H_n^{\mathrm{d}}(m)$ denotes the mth bin of the M-length DFT of a discrete-time equivalent channel $h_n^{\mathrm{d}}(l)$. The scaling factor $\sqrt{M}$ appeared since we are working with the normalized version of the DFT. In general, we assume that the number of subcarriers M employed in CP-OFDM systems is larger than or equal to the length $L+1$ of the channel model, i.e., we assume that $M > L$. This means that we may have to insert some zeros at the end of the channel model $h_n^{\mathrm{d}}(l)$ before computing the M-length DFT $H_n^{\mathrm{d}}(m)$. Such insertion is also known as *zero-padding*.[16]

It is worth mentioning that the notation here is "heavier" than the notation in Chapter 2 since in this chapter we are working with time-varying continuous and discrete variables. So, we shall keep in this chapter the dependency on the index of the OFDM symbol denoted by n and we shall use the superscript "d" for the discrete-time model of the channel $h_n^{\mathrm{d}}(l)$, instead of the simpler notation $h(l)$.

Using the results presented in Subsections 3.3.1, 3.3.2, and 3.3.3, we can now clearly connect the discrete-time implementation of CP-OFDM described in Chapter 2, Subsection 2.4.1 and the CP-OFDM derived through a discretization of the analog OFDM.

3.3.4 BLOCK-BASED MODEL

This section describes a block-based model of the CP-OFDM transceiver, based on the results derived in the previous subsections. Such a model is exactly equal to the model previously presented

[16]This does not have connections with the zero-padding (ZP) type of guard interval that will be explained in Subsection 3.4.2.

in Equations (2.44) and (2.45) of Chapter 2, except for a scaling factor which appears due to the definition of analog OFDM. In that chapter we have proved that those choices of transmitter and receiver matrices are quite useful since they transform the channel matrix into a circulant one, which in turn can be diagonalized using DFT matrices. Such approach is very insightful from the mathematical viewpoint, but it may lack from physical meaning.

Our objective here is to show the structure of the transmitter and receiver matrices, highlighting that such simple structures are natural consequences of the sampling process implemented based on a physically meaningful analog OFDM version.

Firstly, consider the transmitter side. From what we have shown in Subsection 3.3.2, we can form a block of elements $\mathbf{u}(n) \in \mathbb{C}^{N \times 1}$ representing the nth extended OFDM symbol, as follows:

$$\begin{aligned}
\mathbf{u}(n) &\triangleq \begin{bmatrix} u_{n,(-K)} & \cdots & u_{n,(-1)} & u_{n,0} & \cdots & u_{n,(M-1)} \end{bmatrix}^T \\
&= \begin{bmatrix} u_{n,(M-K)} & \cdots & u_{n,(M-1)} & u_{n,0} & \cdots & u_{n,(M-1)} \end{bmatrix}^T \\
&= \underbrace{\begin{bmatrix} \mathbf{0}_{K\times(M-K)} & \mathbf{I}_K \\ \multicolumn{2}{c}{\mathbf{I}_M} \end{bmatrix}}_{\triangleq \mathbf{A}_{\mathrm{CP}}} \begin{bmatrix} u_{n,0} \\ \vdots \\ u_{n,(M-1)} \end{bmatrix} \\
&= \sqrt{\frac{M}{T-\tau}}\mathbf{A}_{\mathrm{CP}} \begin{bmatrix} \mathrm{IDFT}\{s_m(n)\}_0 \\ \vdots \\ \mathrm{IDFT}\{s_m(n)\}_{(M-1)} \end{bmatrix} \\
&= \sqrt{\frac{M}{T-\tau}}\mathbf{A}_{\mathrm{CP}}\mathbf{W}_M^H \underbrace{\begin{bmatrix} s_0(n) \\ \vdots \\ s_{M-1} \end{bmatrix}}_{\triangleq \mathbf{s}(n)} \\
&= \sqrt{\frac{M}{T-\tau}}\mathbf{A}_{\mathrm{CP}}\mathbf{W}_M^H \mathbf{s}(n),
\end{aligned} \tag{3.68}$$

where $\mathbf{W}_M$ is the unitary $M \times M$ DFT matrix defined in Equation (2.46). There is a clear resemblance between the former equation and Equation (2.44) of Chapter 2. Note that this structure appeared in a rather natural way as a consequence of the stacking of the related variables.

Secondly, consider the receiver side. If we define the matrix containing the channel-attenuation factors $\mathbf{\Lambda}(n) \in \mathbb{C}^{M \times M}$ and the vector comprised of the received elements $\mathbf{y}(n) \in \mathbb{C}^{N \times 1}$, with $N = M + K$, as

$$\mathbf{\Lambda}(n) \triangleq \mathrm{diag}\{\lambda_m(n)\}_{m\in\mathcal{M}}, \tag{3.69}$$

$$\mathbf{y}(n) \triangleq \begin{bmatrix} y_{n,(-K)} & \cdots & y_{n,(-1)} & y_{n,0} & \cdots & y_{n,(M-1)} \end{bmatrix}^T, \tag{3.70}$$

then

$$\mathbf{\Lambda}(n)\mathbf{s}(n) = \begin{bmatrix} \lambda_0(n)s_0(n) & \cdots & \lambda_{M-1}(n)s_{M-1}(n) \end{bmatrix}^T, \tag{3.71}$$

and, therefore, we can rewrite Equation (3.49) as follows:

$$\begin{aligned}
\mathbf{\Lambda}(n)\mathbf{s}(n) &= \sqrt{\frac{T-\tau}{M}}\left[\text{DFT}\left\{y_{n,k}\right\}_0 \quad \cdots \quad \text{DFT}\left\{y_{n,k}\right\}_{M-1}\right]^T \\
&= \sqrt{\frac{T-\tau}{M}}\mathbf{W}_M \begin{bmatrix} y_{n,0} \\ \vdots \\ y_{n,(M-1)} \end{bmatrix} \\
&= \sqrt{\frac{T-\tau}{M}}\mathbf{W}_M \underbrace{\left[\mathbf{0}_{M\times K} \quad \mathbf{I}_M\right]}_{\mathbf{R}_{\text{CP}}}\mathbf{y}(n) \\
&= \sqrt{\frac{T-\tau}{M}}\mathbf{W}_M\mathbf{R}_{\text{CP}}\mathbf{y}(n).
\end{aligned} \tag{3.72}$$

Hence, we can define an estimate for $\mathbf{s}(n)$ in the following manner:

$$\hat{\mathbf{s}}(n) = \sqrt{\frac{T-\tau}{M}}\mathbf{E}(n)\mathbf{W}_M\mathbf{R}_{\text{CP}}\mathbf{y}(n), \tag{3.73}$$

where the equalizer matrix $\mathbf{E}(n)$ is an $M \times M$ *diagonal* matrix containing the scaling factors corresponding to each one of the M subcarriers. Once again, it is clear the similarities between the former expression and Equation (2.45) of Chapter 2. In addition, as pointed out in Subsection 3.3.2, one can choose

$$[\mathbf{E}(n)]_{mm} = \frac{1}{\lambda_m(n)}, \tag{3.74}$$

for the ZF solution, or

$$[\mathbf{E}(n)]_{mm} = \frac{\lambda_m^*(n)}{|\lambda_m(n)|^2 + \frac{\sigma_v^2(n)}{\sigma_s^2(n)}}, \tag{3.75}$$

for the MMSE solution (considering the presence of an additive noise signal at the receiver front-end).

As mentioned before, the OFDM discussed here is the so-called *cyclic prefix OFDM* (CP-OFDM), whose block diagram is depicted in Figure 3.10. For the CP-OFDM the equalizer consists of a scalar correction performed at each DFT output to compensate for the channel distortion $\lambda_m(n)$ at the mth frequency bin. The digital channel appearing in Figure 3.10 can also be modeled within the block-based framework by simply rewriting Equation (3.63) as follows (compare with

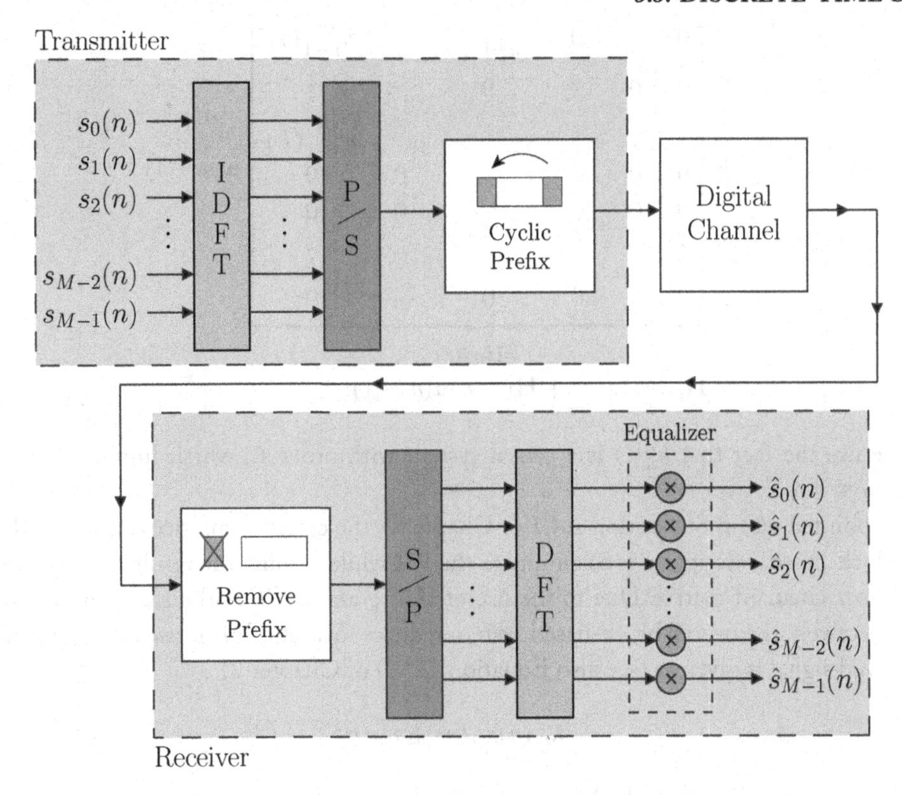

Figure 3.10: OFDM digital transceiver with CP.

Equations (2.38) and (2.39) of Chapter 2):

$$\mathbf{y}(n) = \underbrace{\begin{bmatrix} h_n^{\mathrm{d}}(0) & 0 & 0 & \cdots & 0 \\ h_n^{\mathrm{d}}(1) & h_n^{\mathrm{d}}(0) & 0 & \cdots & 0 \\ \vdots & \vdots & \vdots & \vdots & \vdots \\ h_n^{\mathrm{d}}(L) & h_n^{\mathrm{d}}(L-1) & \ddots & \cdots & 0 \\ 0 & h_n^{\mathrm{d}}(L) & \cdots & \cdots & 0 \\ \vdots & \vdots & \vdots & \vdots & \vdots \\ 0 & 0 & h_n^{\mathrm{d}}(L) & \cdots & h_n^{\mathrm{d}}(0) \end{bmatrix}}_{\triangleq \mathbf{H}_{\mathrm{ISI}}(n)} \mathbf{u}(n)$$

$$+\underbrace{\begin{bmatrix} 0 & \cdots & 0 & h^{\mathrm{d}}_{n-1}(L) & \cdots & h^{\mathrm{d}}_{n-1}(1) \\ 0 & 0 & \cdots & 0 & \ddots & \vdots \\ \vdots & \vdots & \vdots & \vdots & \vdots & h^{\mathrm{d}}_{n-1}(L) \\ 0 & 0 & 0 & \cdots & 0 & 0 \\ 0 & 0 & 0 & \cdots & 0 & 0 \\ \vdots & \vdots & \vdots & \vdots & \vdots & \vdots \\ 0 & 0 & 0 & 0 & \cdots & 0 \end{bmatrix}}_{\triangleq \mathbf{H}_{\mathrm{IBI}}(n)} \mathbf{u}(n-1)$$

$$= \mathbf{H}_{\mathrm{ISI}}(n)\mathbf{u}(n) + \mathbf{H}_{\mathrm{IBI}}(n)\mathbf{u}(n-1), \tag{3.76}$$

where we use the fact that $h^{\mathrm{d}}_n(l)$ is a causal system with order L, which implies that $h^{\mathrm{d}}_n(l) = 0$ whenever $l \in \mathbb{Z} \setminus \mathcal{L}$.

As pointed out in Subsection 2.4.1 of Chapter 2, the effect of matrices $\mathbf{A}_{\mathrm{CP}}$ and $\mathbf{R}_{\mathrm{CP}}$ in the former block-based description is to eliminate the IBI while turning the resulting effective channel matrix into a circulant matrix. Due to the use of IDFT and DFT at the transmitter and receiver sides, respectively, the overall block-based system reduces to a set of M *uncoupled subchannels* whose mth received signal is given as (see also Equation (2.58) of Chapter 2)

$$\lambda_m(n)s_m(n) + v_m(n), \tag{3.77}$$

in which we include the effect of additive noise modeled by $v_m(n)$. The parameter $\lambda_m(n)$ is the so-called *channel tap* associated with the mth subchannel of the nth transmitted block. This uncoupled subchannel model will be very useful in the description of DMT in Subsection 3.4.4.

Example 3.1 (Block-Based Model) Design a CP-OFDM communication system with four subcarriers and a fixed noiseless channel with transfer function

$$H^{\mathrm{d}}_n(z) = 1 + 0.1z^{-1}. \tag{3.78}$$

For simplicity reasons, consider that $T - \tau = M$.

Solution. Firstly, observe that $L = 1$ and $M = 4$. We can therefore choose $K = 1$, yielding $N = M + K = 5$. Hence, the received signals are given as

$$\begin{bmatrix} y_{n,(-1)} \\ y_{n,0} \\ y_{n,1} \\ y_{n,2} \\ y_{n,3} \end{bmatrix} = \begin{bmatrix} 1 & 0 & 0 & 0 & 0 \\ 0.1 & 1 & 0 & 0 & 0 \\ 0 & 0.1 & 1 & 0 & 0 \\ 0 & 0 & 0.1 & 1 & 0 \\ 0 & 0 & 0 & 0.1 & 1 \end{bmatrix} \begin{bmatrix} u_{n,(-1)} \\ u_{n,0} \\ u_{n,1} \\ u_{n,2} \\ u_{n,3} \end{bmatrix} + \begin{bmatrix} 0 & 0 & 0 & 0 & 0.1 \\ 0 & 0 & 0 & 0 & 0 \\ 0 & 0 & 0 & 0 & 0 \\ 0 & 0 & 0 & 0 & 0 \\ 0 & 0 & 0 & 0 & 0 \end{bmatrix} \begin{bmatrix} u_{(n-1),(-1)} \\ u_{(n-1),0} \\ u_{(n-1),1} \\ u_{(n-1),2} \\ u_{(n-1),3} \end{bmatrix}$$

$$= \begin{bmatrix} 1 & 0 & 0 & 0 & 0 \\ 0.1 & 1 & 0 & 0 & 0 \\ 0 & 0.1 & 1 & 0 & 0 \\ 0 & 0 & 0.1 & 1 & 0 \\ 0 & 0 & 0 & 0.1 & 1 \end{bmatrix} \begin{bmatrix} \mathbf{0}_{1\times 3} & 1 \\ \multicolumn{2}{c}{\mathbf{I}_4} \end{bmatrix} \begin{bmatrix} u_{n,0} \\ u_{n,1} \\ u_{n,2} \\ u_{n,3} \end{bmatrix}$$

$$+ \begin{bmatrix} 0 & 0 & 0 & 0 & 0.1 \\ 0 & 0 & 0 & 0 & 0 \\ 0 & 0 & 0 & 0 & 0 \\ 0 & 0 & 0 & 0 & 0 \\ 0 & 0 & 0 & 0 & 0 \end{bmatrix} \begin{bmatrix} \mathbf{0}_{1\times 3} & 1 \\ \multicolumn{2}{c}{\mathbf{I}_4} \end{bmatrix} \begin{bmatrix} u_{(n-1),0} \\ u_{(n-1),1} \\ u_{(n-1),2} \\ u_{(n-1),3} \end{bmatrix}$$

$$= \begin{bmatrix} 0 & 0 & 0 & 1 \\ 1 & 0 & 0 & 0.1 \\ 0.1 & 1 & 0 & 0 \\ 0 & 0.1 & 1 & 0 \\ 0 & 0 & 0.1 & 1 \end{bmatrix} \begin{bmatrix} u_{n,0} \\ u_{n,1} \\ u_{n,2} \\ u_{n,3} \end{bmatrix} + \begin{bmatrix} 0 & 0 & 0 & 0.1 \\ 0 & 0 & 0 & 0 \\ 0 & 0 & 0 & 0 \\ 0 & 0 & 0 & 0 \\ 0 & 0 & 0 & 0 \end{bmatrix} \begin{bmatrix} u_{(n-1),0} \\ u_{(n-1),1} \\ u_{(n-1),2} \\ u_{(n-1),3} \end{bmatrix}. \tag{3.79}$$

The first operation at the receiver is the removal of the redundancy through the multiplication by $\mathbf{R}_{\text{CP}} = [\mathbf{0}_{4\times 1} \;\; \mathbf{I}_4]$. This operation is equivalent to discarding the first element, $y_{n,(-1)}$, of the received data block, thus removing the effect of the IBI and yielding the following description:

$$\begin{bmatrix} y_{n,0} \\ y_{n,1} \\ y_{n,2} \\ y_{n,3} \end{bmatrix} = \begin{bmatrix} 1 & 0 & 0 & 0.1 \\ 0.1 & 1 & 0 & 0 \\ 0 & 0.1 & 1 & 0 \\ 0 & 0 & 0.1 & 1 \end{bmatrix} \begin{bmatrix} u_{n,0} \\ u_{n,1} \\ u_{n,2} \\ u_{n,3} \end{bmatrix}$$

$$= \begin{bmatrix} 1 & 0 & 0 & 0.1 \\ 0.1 & 1 & 0 & 0 \\ 0 & 0.1 & 1 & 0 \\ 0 & 0 & 0.1 & 1 \end{bmatrix} \mathbf{W}_4^H \begin{bmatrix} s_0(n) \\ s_1(n) \\ s_2(n) \\ s_3(n) \end{bmatrix}$$

$$= \frac{1}{2} \begin{bmatrix} 1 & 0 & 0 & 0.1 \\ 0.1 & 1 & 0 & 0 \\ 0 & 0.1 & 1 & 0 \\ 0 & 0 & 0.1 & 1 \end{bmatrix} \begin{bmatrix} 1 & 1 & 1 & 1 \\ 1 & \mathrm{j} & -1 & -\mathrm{j} \\ 1 & -1 & 1 & -1 \\ 1 & -\mathrm{j} & -1 & \mathrm{j} \end{bmatrix} \begin{bmatrix} s_0(n) \\ s_1(n) \\ s_2(n) \\ s_3(n) \end{bmatrix}. \tag{3.80}$$

Now, by taking the DFT of the resulting signal, one has

$$\mathbf{W}_4 \begin{bmatrix} y_{n,0} \\ y_{n,1} \\ y_{n,2} \\ y_{n,3} \end{bmatrix} = \frac{1}{4} \begin{bmatrix} 1 & 1 & 1 & 1 \\ 1 & -\mathrm{j} & -1 & \mathrm{j} \\ 1 & -1 & 1 & -1 \\ 1 & \mathrm{j} & -1 & -\mathrm{j} \end{bmatrix} \begin{bmatrix} 1 & 0 & 0 & 0.1 \\ 0.1 & 1 & 0 & 0 \\ 0 & 0.1 & 1 & 0 \\ 0 & 0 & 0.1 & 1 \end{bmatrix} \begin{bmatrix} 1 & 1 & 1 & 1 \\ 1 & \mathrm{j} & -1 & -\mathrm{j} \\ 1 & -1 & 1 & -1 \\ 1 & -\mathrm{j} & -1 & \mathrm{j} \end{bmatrix} \begin{bmatrix} s_0(n) \\ s_1(n) \\ s_2(n) \\ s_3(n) \end{bmatrix}$$
$$= \begin{bmatrix} 1.1 & 0 & 0 & 0 \\ 0 & 1-0.1\mathrm{j} & 0 & 0 \\ 0 & 0 & 0.9 & 0 \\ 0 & 0 & 0 & 1+0.1\mathrm{j} \end{bmatrix} \begin{bmatrix} s_0(n) \\ s_1(n) \\ s_2(n) \\ s_3(n) \end{bmatrix}, \tag{3.81}$$

leading to an uncoupled model with four subchannels. □

3.4 OTHER OFDM-BASED SYSTEMS

In practice, there are several distinct ways of configuring the block transceivers that are closely related to the CP-OFDM discussed in Section 3.3. In this section, we shall present some of them, starting from the single-carrier with frequency-domain equalization (SC-FD) in Subsection 3.4.1. This type of transceiver is able to significantly reduce the PAPR as compared to the CP-OFDM, a very desirable feature from the implementation viewpoint. In addition, we will also describe the so-called ZP-OFDM in Subsection 3.4.2, which employs zero-valued elements in the guard interval, instead of a cyclic prefix. This saves transmitter power, a valuable resource in many applications. Another very useful system in practice is the so-called coded OFDM (C-OFDM), which combines the good features inherent to OFDM system with the additional protection enabled by error-correcting codes. In this context, Subsection 3.4.3 gives a brief introduction to C-OFDM systems. The section ends with Subsection 3.4.4, which describes a standard manner of taking into account the knowledge of the channel state at the transmitter side in order to enhance the related data rates. The discrete multitone (DMT) systems are able to take into account such side information about the channel by using a technique called water-filling.

3.4.1 SC-FD

As previously discussed in Subsection 3.2.4, the OFDM symbol presents high PAPR imposing a high linearity constraint in the transmitter power amplifier. In the case where the power of the transmitted signal has high dynamic range, it is possible that the nonlinear behavior of the transmitter amplifier introduces distortions to the transmitted OFDM symbol, leading to loss of orthogonality among the subcarriers. The *deterministic PAPR* of a discrete-time OFDM symbol can be defined as

$$\mathrm{PAPR}_{\mathrm{OFDM}}(n) \triangleq \frac{\max_k \left\{ |u_{n,k}|^2 \right\}}{\frac{1}{M} \sum_{k \in \mathcal{M}} |u_{n,k}|^2}, \tag{3.82}$$

in which $\max_k \left\{|u_{n,k}|^2\right\}$ denotes the peak power level associated with the nth OFDM symbol, whereas $\frac{1}{M}\sum_{k\in\mathcal{M}} |u_{n,k}|^2$ denotes the average power of this nth block. In the case of CP-OFDM, the insertion of the guard interval does not change the maximum achievable power (peak) of the elements that compose an OFDM symbol. In addition, it is very unlikely that the guard interval could change the average power as well. Therefore, we can consider only the elements after the application of the IDFT, without considering the cyclic prefix which is inserted. Thus, based on Equation (3.55), one has

$$u_{n,k} = \frac{1}{\sqrt{T-\tau}} \sum_{m\in\mathcal{M}} s_m(n) e^{j\frac{2\pi}{M}mk}, \tag{3.83}$$

yielding (see also Equation (3.42))

$$\begin{aligned}
\frac{1}{M}\sum_{k\in\mathcal{M}} |u_{n,k}|^2 &= \frac{1}{M}\sum_{k\in\mathcal{M}} u_{n,k}u_{n,k}^* \\
&= \frac{1}{M}\sum_{k\in\mathcal{M}} \left[\frac{1}{T-\tau}\sum_{m\in\mathcal{M}}\sum_{m'\in\mathcal{M}} s_m(n)s_{m'}^*(n)e^{j\frac{2\pi}{M}(m-m')k}\right] \\
&= \frac{1}{T-\tau}\sum_{m\in\mathcal{M}}\sum_{m'\in\mathcal{M}} s_m(n)s_{m'}^*(n)\left[\frac{1}{M}\sum_{k\in\mathcal{M}} e^{j\frac{2\pi}{M}(m-m')k}\right] \\
&= \frac{1}{T-\tau}\sum_{m\in\mathcal{M}} s_m(n)\sum_{m'\in\mathcal{M}} s_{m'}^*(n)\delta[m-m'] \\
&= \frac{1}{T-\tau}\sum_{m\in\mathcal{M}} |s_m(n)|^2 \\
&= \frac{M}{T-\tau}\left[\frac{1}{M}\sum_{m\in\mathcal{M}} |s_m(n)|^2\right],
\end{aligned} \tag{3.84}$$

and

$$\begin{aligned}
|u_{n,k}| &= \frac{1}{\sqrt{T-\tau}}\left|\sum_{m\in\mathcal{M}} s_m(n)e^{j\frac{2\pi}{M}mk}\right| \\
&\leq \frac{1}{\sqrt{T-\tau}}\sum_{m\in\mathcal{M}} \left|s_m(n)e^{j\frac{2\pi}{M}mk}\right| \\
&= \frac{1}{\sqrt{T-\tau}}\sum_{m\in\mathcal{M}} |s_m(n)| \\
&\leq \frac{1}{\sqrt{T-\tau}}\sum_{m\in\mathcal{M}} \max_m\{|s_m(n)|\} \\
&= \frac{M}{\sqrt{T-\tau}}\max_m\{|s_m(n)|\},
\end{aligned} \tag{3.85}$$

thus implying that

$$\max_k |u_{n,k}|^2 \leq \frac{M^2}{T-\tau} \max_m \left\{ |s_m(n)|^2 \right\}. \tag{3.86}$$

The results of Equations (3.84) and (3.86) give us that

$$\begin{aligned} \mathrm{PAPR}_{\mathrm{OFDM}}(n) &\leq M \underbrace{\left(\frac{\max\limits_m \left\{ |s_m(n)|^2 \right\}}{\frac{1}{M} \sum\limits_{m \in \mathcal{M}} |s_m(n)|^2} \right)}_{\triangleq \mathrm{PAPR}_{\mathrm{SC}}(n)} \\ &= M \mathrm{PAPR}_{\mathrm{SC}}(n), \end{aligned} \tag{3.87}$$

where SC stands for single-carrier transmissions, i.e., no pre-processing is implemented at the transmitter side. The result of Equation (3.87) means that the upper bound for the PAPR related to OFDM systems is M times larger than the corresponding PAPR of a single-carrier system. In practice, it is observed that this is a tight upper bound, which means that the actual value of $\mathrm{PAPR}_{\mathrm{OFDM}}(n)$ is close to $M\mathrm{PAPR}_{\mathrm{SC}}(n)$. In many applications, M is made large in order to cope with the loss of spectral efficiency due to the redundancy insertion (guard interval), which might turn the PAPR problem even more critical [45].[17]

A possible solution to the PAPR issue is to perform the block transmission in a single-carrier fashion while performing the equalization in the frequency domain, as illustrated in Figure 3.11. In this configuration the cyclic prefix is inserted at the transmitter end, whereas the IDFT is implemented at the receiver in order to allow a decoupled equalization, which is represented by the multipliers between the IDFT and DFT blocks in Figure 3.11. This block transmission setup is called *single-carrier with frequency-domain equalization* (SC-FD) transceiver. Since it uses a cyclic prefix, we will call it cyclic prefix SC-FD (CP-SC-FD), as also described in Subsection 2.4.3.

3.4.2 ZP-BASED SCHEMES

The zero-padding (ZP) represents an alternative to the use of CP as guard time. ZP is a quite effective way to eliminate the IBI that pervades block-based transmissions. Indeed, in several different setups, ZP systems are optimal solutions in the MSE sense [83]. This optimality characteristic leads to better performance of ZP-based transceivers, as compared to CP-based systems in a number of situations [55].

In comparison with CP-based schemes, one of the main advantages of ZP-based transceivers is their lower required power due to their trailing zeros. On the other hand, ZP-based transceivers are usually able to eliminate only IBI effects, whereas the ISI effects should be mitigated by a

[17] Another definition for the PAPR considers the OFDM symbol as a random sequence and takes the expected value of that sequence instead of the deterministic average value. However, the results and conclusions are the same of those presented here. See, for instance, [45] for further details.

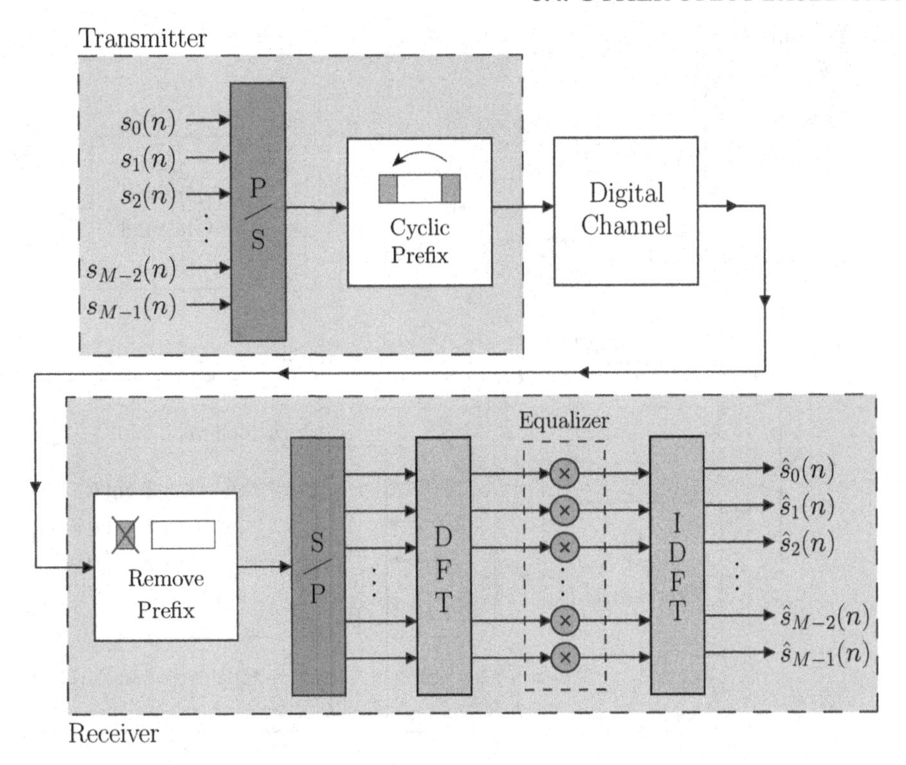

Figure 3.11: SC-FD digital transceiver with CP.

properly designed equalizer. Thus, in general, ZP-based transceivers do not yield so simple (one-tap) equalizers as CP-based systems do. These observations are valid for the general ZP versions of the OFDM and SC-FD, herein called ZP-OFDM and ZP-SC-FD, respectively. In fact, there are some tools related to structured matrix representations which can work around these difficulties, yielding general ZP-transceivers which are still based on DFTs and one-tap equalizers (see Figure 4.3 of Chapter 4). However, there are particular versions of ZP-OFDM and ZP-SC-FD that have the same simple equalizers as the ones used in CP-OFDM and CP-SC-FD systems. These particular versions, known as ZP-OFDM-OLA and ZP-SC-FD-OLA, perform overlap-and-add (OLA) operations at the receiver side and they are depicted in Figures 3.12 and 3.13. The mathematical details regarding these particular versions of ZP-OFDM and ZP-SC-FD were already explained in Subsections 2.4.2 and 2.4.3, respectively.

The following topics are research results concerning the differences between the general ZP-OFDM and ZP-SC-FD and their cyclic-prefix counterparts [55, 87]:

- ZP-OFDM introduces more nonlinear distortion than CP-OFDM transceivers.

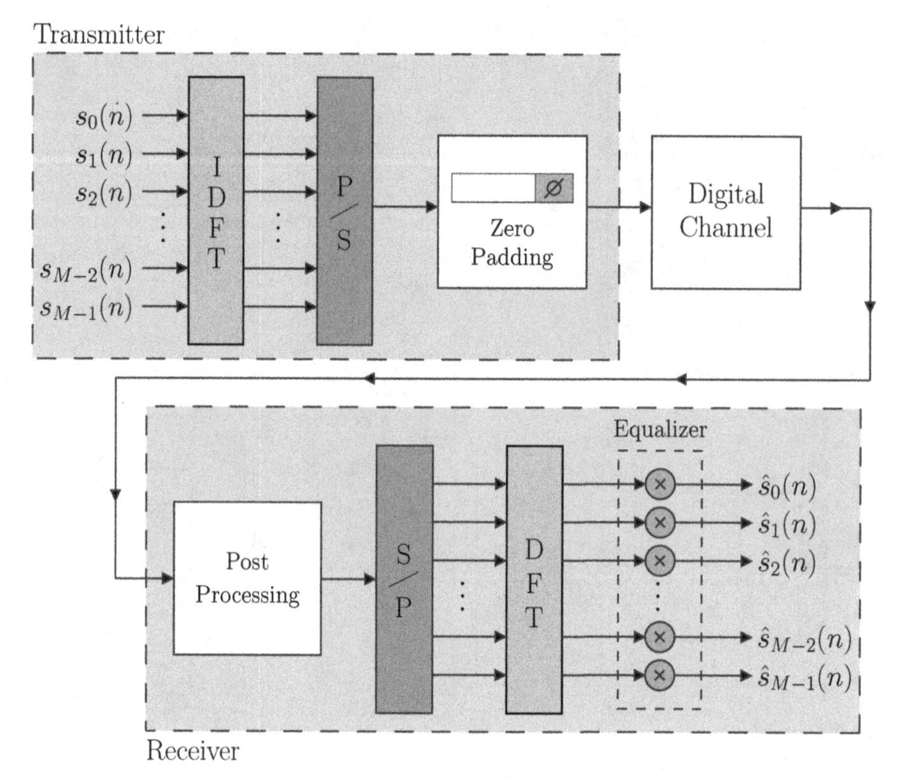

Figure 3.12: OFDM digital transceiver with ZP.

- ZP-OFDM has better performance in terms of BER or MSE than CP-OFDM, for a given average-bit-energy-to-noise power ratio, E_b/N_0.
- The ZP-SC-FD has lower PAPR, presents robustness to CFO, and has also better uncoded performance. However, the equalization is a bit more complex to implement.
- In the case some kind of channel coding is included (C-OFDM), the coded version is better when code rate is low and the error correcting coding capability is enhanced. In the coded case ZP-SC-FD is better than C-OFDM for high code rate.
- ZP-SC-FD only has clear performance advantages over uncoded OFDM.
- Uncoded CP-OFDM is inferior to zero-forcing equalized CP-SC-FD transceiver.
- CP-OFDM with equal-gain power allocation has the same performance as zero-forcing equalized CP-SC-FD transceiver.

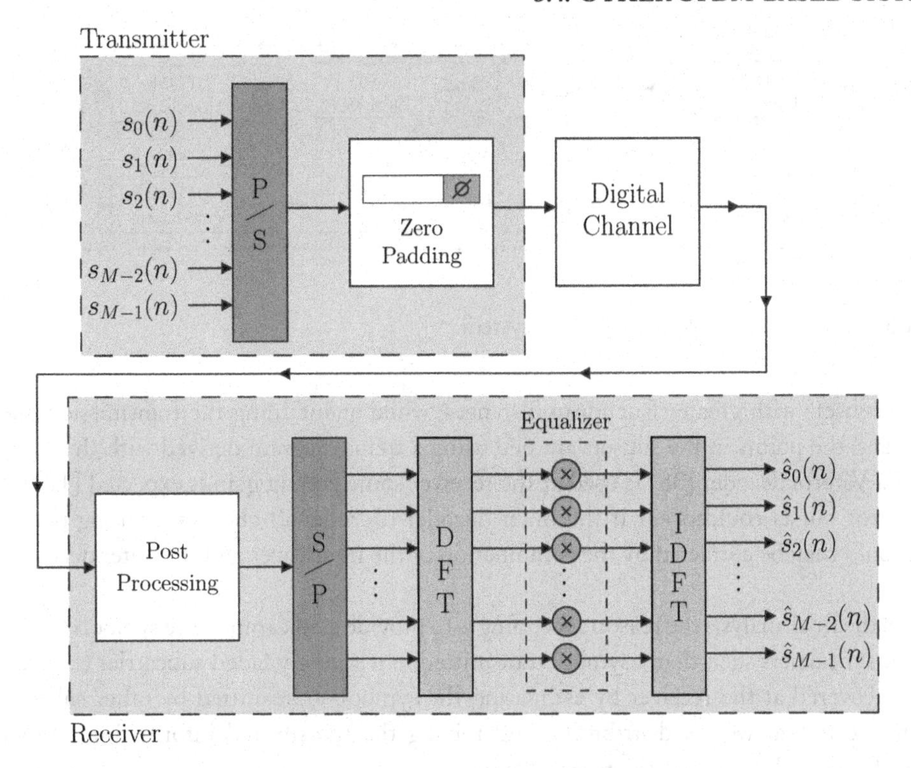

Figure 3.13: SC-FD digital transceiver with ZP.

- CP-OFDM with AMBER (an approximately minimum BER) power allocation is better than zero-forcing equalized CP-SC-FD.

3.4.3 CODED OFDM

The usual complete coded OFDM (C-OFDM) system utilized in some broadcast systems is depicted in Figure 3.14 where two levels of coding and an interleaving are employed. These building blocks are required to protect the transmitted information against distortions and deep fadings in some subcarriers [13, 20]. C-OFDM includes a few building blocks as described in [13].

The outer encoder (coder 1) is meant to insert redundancy in the data stream, e.g., using Reed-Solomon codes [38, 90]. This will increase the required bandwidth for transmission, but on the other hand, allows more reliable reception. The aim is either to obtain block codes with increased Hamming distances or to encode the data in a continuous way such that a correlation between the data is induced to help detection at the receiver end.

The inner encoder (coder 2) is meant to increase the Euclidean distance among the symbols of the constellation, usually employing a trellis coded modulation. The idea is to increase the number of possible symbols in comparison with the number of points in the given constellation and subdivide

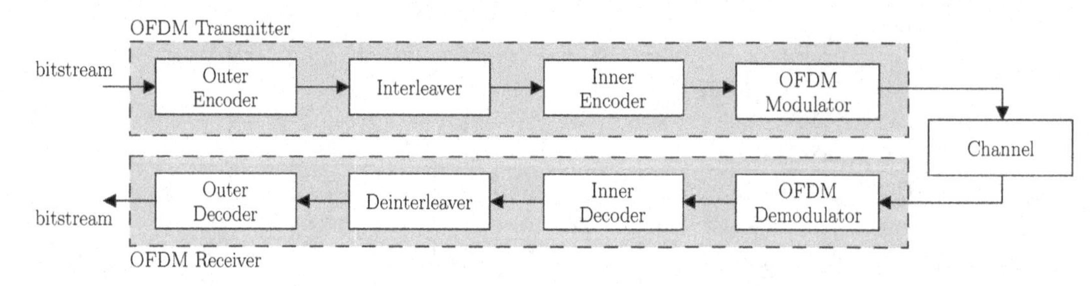

Figure 3.14: Coded OFDM (C-OFDM) system.

them in subsets with greater Euclidean distances, while maintaining the transmission energy. The subsets and the points in the subsets are tied using a trellis diagram derived with the convolutional code. If a Viterbi decoder [38] is used at the receiver some coding gain is expected [13].

Error bursts could occur if the inner decoder (decoder 2) chooses a wrong decoding path which could then be corrected by the combination of the interleaver and the outer decoder (decoder 1).

In the final analysis the reason for coding is to provide a link among the symbols transmitted on different subcarriers such that a symbol transmitted in a strongly faded subcarrier (i.e., $\lambda_m(n) \approx 0$) can be recovered at the receiver by estimating the symbols transmitted by other subcarriers. The bottom line is that we are distributing and mixing the transmitted information to increase the chance of proper detection at the receiver end.

3.4.4 DMT

A *discrete multitone* (DMT) transceiver is essentially an OFDM system comprised of three particular features: (i) there is no passband conversion to a higher carrier frequency, which means that the actual transmitted signals are baseband signals; (ii) since any actual transmission employs real-valued signals, then the baseband transmitted signals must be real-valued. This means that the input constellation symbols $s_m(n)$ must have the conjugate symmetric property, i.e., $s_m(n) = s^*_{(M-m)}(n)$, for all $m \in \mathcal{M}$; and (iii) there is some kind of channel-state information (CSI) at the transmitter side, so that the transceiver can use some smart techniques in order to cope with possible channel impairments in advance.

The third DMT property above is indeed its key feature since it enables transmissions with higher data rates. Nevertheless, it is usually applicable in wired connections, in which the channel state does not change too often. As described in Subsection 1.3.1 of Chapter 1, the DMT system is currently employed in many digital subscriber line (xDSL) applications. The aim of this subsection is to describe how DMT-based systems take into account the availability of information about the channel at the transmitter side in order to enhance the overall transmission performance.

As described in Subsection 3.3.4, OFDM systems can be thought as M parallel uncoupled subchannels, whose mth received signal of the nth transmitted OFDM symbol is given by

$\lambda_m(n)s_m(n) + v_m(n)$, in which we consider that $s_m(n)$ is a random sequence with zero-mean and variance $\sigma^2_{s,m}(n)$, whereas $v_m(n)$ is a random sequence with zero-mean and variance $\sigma^2_{v,m}(n)$, where $m \in \mathcal{M}$ and $n \in \mathbb{Z}$. In this case, we have the following signal-to-noise ratio (SNR) associated with the mth subchannel:

$$\text{SNR}_m(n) \triangleq \frac{|\lambda_m(n)|^2 \sigma^2_{s,m}(n)}{\sigma^2_{v,m}(n)}. \tag{3.88}$$

If $v_m(n)$ is a Gaussian random sequence, then it follows from basic knowledge on digital communications that, for each fixed pair of numbers $(m, n) \in \mathcal{M} \times \mathbb{Z}$, one has the following error probability of symbols [22, 26, 45, 66, 83]:

$$P_{\text{e},m}(n) \approx c\mathcal{Q}\left(A\sqrt{\text{SNR}_m(n)}\right), \tag{3.89}$$

in which $\mathcal{Q}(\cdot)$ is the area under a Gaussian tail, being a decreasing function of its argument defined as

$$\mathbb{R} \ni x \mapsto \mathcal{Q}(x) \triangleq \frac{1}{\sqrt{2\pi}} \int_x^{\infty} \mathrm{e}^{-w^2/2} \mathrm{d}w, \tag{3.90}$$

whereas c and A depend upon the number of bits $b_m(n) \in \mathbb{N}$ employed to represent $s_m(n)$. Large number of bits leads to both relatively large values of c and small values of A. In the particular case of a $2^{b_m(n)}$-QAM constellation, for instance, one has [83]

$$c = 4(1 - 2^{-\frac{b_m(n)}{2}}), \tag{3.91}$$

$$A = \sqrt{\frac{3}{2^{b_m(n)} - 1}}. \tag{3.92}$$

This means that, for $s_m(n)$ pertaining to a $2^{b_m(n)}$-QAM constellation, one can rewrite Equation (3.89) as follows:

$$A \approx \frac{1}{\sqrt{\text{SNR}_m(n)}} \mathcal{Q}^{-1}\left(\frac{P_{\text{e},m}(n)}{c}\right), \tag{3.93}$$

yielding (see Equations (3.91) and (3.92))

$$\begin{aligned} b_m(n) &\approx \log_2\left(1 + \frac{\text{SNR}_m(n)}{\left[\mathcal{Q}^{-1}\left(\frac{P_{\text{e},m}(n)}{c}\right)\right]^2/3}\right) \\ &= \log_2\left(1 + \frac{\text{SNR}_m(n)}{\Gamma_{\text{QAM}}}\right), \end{aligned} \tag{3.94}$$

where the number Γ_{QAM} is the so-called *SNR gap* associated with the QAM constellation, being defined as

$$\Gamma_{\text{QAM}} \triangleq \frac{\left[\mathcal{Q}^{-1}\left(\frac{P_{\text{e},m}(n)}{c}\right)\right]^2}{3}. \tag{3.95}$$

Strictly speaking, Equation (3.94) is not a closed-form expression for $b_m(n)$ since the definition of c in Equation (3.92) depends on $b_m(n)$, thus implying that the SNR gap is a function of the number of bits itself. Nevertheless, the dependency of Γ_{QAM} on $b_m(n)$ is quite weak in the sense that Γ_{QAM} varies slowly when the number of bits changes. For instance, considering an error probability of symbols of $P_{\text{e},m}(n) = 10^{-5}$, then $\Gamma_{\text{QAM}} \in [6.5, 6.9]$ when $b_m(n)$ varies from 2 to 8. Besides, when $b_m(n)$ is large, then $\Gamma_{\text{QAM}} \approx \left[\mathcal{Q}^{-1}\left(P_{\text{e},m}(n)/4\right)\right]^2/3$, i.e., there is no dependence on $b_m(n)$.

The important conclusion that we can draw from Equation (3.94) is that power and bit allocations are closely related problems. Indeed, for a predefined target error probability of symbols, large values of SNR allow increasing the number of bits following the rule expressed in Equation (3.94). Increasing the number of transmitted bits without sacrificing the error-probability performance is a rather desirable feature of a transmission scheme. On the other hand, we cannot increase the SNR as we wish since this may entail an unrealistic transmission power.

So far, we are working only with one predefined subchannel index $m \in \mathcal{M}$. When all subchannels are considered, then we can work with the average number of bits $b(n) \in \mathbb{R}$ in an OFDM symbol, which is the sum of the number of bits transmitted on each subcarrier divided by the total number of subcarriers M, i.e.,

$$\begin{aligned} b(n) &\approx \frac{1}{M}\sum_{m\in\mathcal{M}} \log_2\left(1+\frac{\text{SNR}_m(n)}{\Gamma_{\text{QAM}}}\right) \\ &= \log_2\left[\prod_{m\in\mathcal{M}}\left(1+\frac{\text{SNR}_m(n)}{\Gamma_{\text{QAM}}}\right)\right]^{\frac{1}{M}} \\ &= \log_2\left(1+\frac{\text{SNR}(n)}{\Gamma_{\text{QAM}}}\right), \end{aligned} \tag{3.96}$$

where $\text{SNR}(n)$ is defined as

$$\text{SNR}(n) \triangleq \prod_{m\in\mathcal{M}} \left[\Gamma_{\text{QAM}} + \text{SNR}_m(n)\right]^{\frac{1}{M}} - \Gamma_{\text{QAM}}, \tag{3.97}$$

which means that $\text{SNR}(n)$ is basically the geometric mean of the sum of the SNR gap with the mth subchannel SNR, subtracting the SNR gap from the final result.

Thus, the parallel and independent subchannels together behave as a single white Gaussian noise channel with SNR corresponding to $\text{SNR}(n)$, which is essentially equal to the *geometric mean* of the subcarrier SNRs increased by the SNR gap. This geometric mean is a good approximation for moderate to high SNRs on all subcarriers. As the geometric mean is much more sensitive to small values than the arithmetic mean,[18] then the existence of poor subchannels (i.e., subchannels with low SNRs) can significantly degrade the system performance in terms of error probability of symbols. An alternative view is that, for a given error-probability performance, the presence of poor subcarriers can significantly decrease the transmission data rates.

[18] In fact, the geometric mean of positive real numbers is always smaller than or equal to the arithmetic mean of these numbers.

The former reasoning clarifies how the quality of the subchannels affects the overall system performance, since the SNR of the mth subchannel is affected directly by the channel gain $\lambda_m(n)$ (see Equation (3.88)). Standard uncoded OFDM-based systems suffer critically from performance degradations caused by deep fadings in some subchannels. In order to appreciate the advantages of the DMT over standard OFDM-based systems, the $\text{SNR}(n)$ must be compared with unbiased SNR of equalized baseband systems. In practice, $\text{SNR}(n)$ can be improved considerably when the available energy is distributed *non-uniformly* over all, or a subset of parallel carriers, giving rise to high-performance multicarrier systems. This energy allocation is implemented by the so-called *loading algorithms*, whose most notorious example is the *water-filling* approach. Let us formulate this approach from now on.

From information theory, it is well known that the ultimate transmission data rate is dictated by the *channel capacity* [14]. Indeed, the channel capacity is the upper bound on the possible data rates which any communication system is able to achieve, assuring that error probability of symbols is zero. It is possible to show that the channel capacity (in bits/transmission, i.e., channel capacity normalized by the channel bandwidth) of the mth Gaussian subchannel of a CP-OFDM system is given by (in the case of complex-valued variables) [14]

$$C_m(n) = \log_2\left(1 + \text{SNR}_m(n)\right), \tag{3.98}$$

implying that the average channel capacity, $C(n)$, considering all M subchannels of the nth data block, is

$$C(n) = \frac{1}{M} \sum_{m \in \mathcal{M}} \log_2\left(1 + \text{SNR}_m(n)\right). \tag{3.99}$$

This expression is quite similar to the average amount of transmitted bits expressed in Equation (3.96).

Therefore, in order to increase the transmission data rates one could try to compensate for the channel taps $\lambda_m(n)$ leading to low the subchannel SNRs, and an intuitive solution to this problem is maximizing the average channel capacity in expression (3.99) subject to a predefined power constraint. It is worth mentioning at this point that, although a transmission scheme that achieves the related capacity could be theoretically implemented by using codes with a long block length [14], practical communications systems avoid high complexity codes, thus implying that the capacity may not be achievable. In other words, for a given small average error probability of symbols $P_e(n)$, the channel code might require infinite complexity to achieve the capacity. In practice, at a given fixed error probability of symbols, one can characterize the modulation/coding by a gap $\Gamma \in \mathbb{R}_+$ (for example, $\Gamma = \Gamma_{\text{QAM}}$ in Equation (3.95)) that quantifies the effective loss in SNR with respect to capacity. Extremely powerful codes (such as turbo code and concatenated code) can reduce the gap to up 1 or 2 dB.

Let us formulate the problem of maximizing the average channel capacity (or minimizing $-C(n)$ expressed using the natural logarithm) as the following optimization problem:

$$\min_{\substack{\sigma_{s,m}(n)\in\mathbb{R}\\ m\in\mathcal{M}}} \left\{ -\frac{1}{M}\sum_{m\in\mathcal{M}} \ln\left(1+\frac{|\lambda_m(n)|^2\sigma_{s,m}^2(n)}{\sigma_{v,m}^2(n)}\right)\right\}, \tag{3.100}$$

subject to:

$$\sum_{m\in\mathcal{M}} \sigma_{s,m}^2(n) = p_0(n), \tag{3.101}$$

in which the positive real number $p_0(n)$ denotes the *total transmission power* of the nth block. In this optimization problem, it is assumed that $\sigma_{v,m}^2(n)$ and $\lambda_m(n)$ are fixed and known numbers, for all $m \in \mathcal{M}$.

By applying the Lagrange-multiplier method (with Lagrange multiplier $\lambda \in \mathbb{R}$), we have the following cost-function:

$$\begin{aligned} J\left(\sigma_{s,0}(n),\cdots,\sigma_{s,(M-1)}(n)\right) \triangleq & -\frac{1}{M}\sum_{m\in\mathcal{M}} \ln\left(1+\frac{|\lambda_m(n)|^2\sigma_{s,m}^2(n)}{\sigma_{v,m}^2(n)}\right) \\ & + \lambda\left(\sum_{m\in\mathcal{M}} \sigma_{s,m}^2(n) - p_0(n)\right), \end{aligned} \tag{3.102}$$

which can be optimized by finding its associated extreme points, as follows:

$$\frac{\partial J\left(\sigma_{s,0}(n),\cdots,\sigma_{s,(M-1)}(n)\right)}{\partial\sigma_{s,m}(n)} = -\frac{2}{M}\frac{\sigma_{s,m}(n)}{\frac{\sigma_{v,m}^2(n)}{|\lambda_m(n)|^2}+\sigma_{s,m}^2(n)} + 2\lambda\sigma_{s,m}(n). \tag{3.103}$$

Thus, for each $m \in \mathcal{M}$, we have

$$\frac{\partial J\left(\sigma_{s,0}(n),\cdots,\sigma_{s,m}^{o}(n),\cdots,\sigma_{s,(M-1)}(n)\right)}{\partial\sigma_{s,m}(n)} = 0 \tag{3.104}$$

if and only if the optimal value $\sigma_{s,m}^{o}(n)$ is such that $\sigma_{s,m}^{o}(n) = 0$ or (discarding the negative root)

$$\sigma_{s,m}^{o}(n) = \sqrt{\frac{1}{M\lambda} - \frac{\sigma_{v,m}^2(n)}{|\lambda_m(n)|^2}}. \tag{3.105}$$

Note that, based on Equation (3.103), when $\sigma_{s,m}^{o}(n) \neq 0$, one has $\lambda \neq 0$, which means that $\frac{1}{\lambda}$ is well defined.[19]

Now, we can consider, without loss of generality, that the numbers $\sigma_{s,m}^{o}(n)$, with $m \in \mathcal{M}$, are ordered as follows:

$$\sigma_{s,0}^{o}(n) \geq \sigma_{s,1}^{o}(n)\cdots \geq \sigma_{s,(M-1)}^{o}(n). \tag{3.106}$$

[19]We shall omit the proof that those optimal values correspond to solutions that minimize the original objective function.

In addition, let us assume that we will transmit using M' subchannels, where $M'-1 \in \mathcal{M}$ is the largest index of the smallest value $\sigma^{\text{o}}_{s,m}(n)$, with $m \in \mathcal{M}$. Note that $M' \leq M$ and that

$$\sigma^{\text{o}}_{s,0}(n) \geq \cdots \geq \sigma^{\text{o}}_{s,(M'-1)}(n) > 0. \tag{3.107}$$

Thus, by defining $\mathcal{M}' \triangleq \{0, \cdots, M'-1\}$, the constraint in Equation (3.101) can be used to find the Lagrange multiplier, as follows:

$$\begin{aligned}\sum_{m' \in \mathcal{M}'} \left(\sigma^{\text{o}}_{s,m'}(n)\right)^2 &= \sum_{m' \in \mathcal{M}'} \left(\frac{1}{M\lambda} - \frac{\sigma^2_{v,m'}(n)}{|\lambda_{m'}(n)|^2}\right) \\ &= \frac{M'}{M}\frac{1}{\lambda} - \sum_{m' \in \mathcal{M}'} \frac{\sigma^2_{v,m'}(n)}{|\lambda_{m'}(n)|^2} \\ &= p_0(n),\end{aligned} \tag{3.108}$$

yielding

$$\frac{1}{M\lambda} = \frac{1}{M'}\left(p_0(n) + \sum_{m' \in \mathcal{M}'} \frac{\sigma^2_{v,m'}(n)}{|\lambda_{m'}(n)|^2}\right). \tag{3.109}$$

Therefore, according to Equation (3.105), the variances of the first $M' \leq M$ constellation symbols of the nth block are given by

$$\left(\sigma^{\text{o}}_{s,m}(n)\right)^2 = \frac{1}{M'}\left(p_0(n) + \sum_{m' \in \mathcal{M}'} \frac{\sigma^2_{v,m'}(n)}{|\lambda_{m'}(n)|^2}\right) - \frac{\sigma^2_{v,m}(n)}{|\lambda_m(n)|^2}, \tag{3.110}$$

with $m \in \mathcal{M}'$, whereas the remaining $M - M'$ symbols do not transmit data, i.e., $\left(\sigma^{\text{o}}_{s,m}(n)\right)^2 = 0$ for $m \in \mathcal{M} \setminus \mathcal{M}'$. In other words, we can state that

$$\left(\sigma^{\text{o}}_{s,m}(n)\right)^2 = \begin{cases} \dfrac{\left(p_0(n) + \sum_{m' \in \mathcal{M}'} \frac{\sigma^2_{v,m'}(n)}{|\lambda_{m'}(n)|^2}\right)}{M'} - \dfrac{\sigma^2_{v,m}(n)}{|\lambda_m(n)|^2}, & \text{if this quantity is larger than zero,} \\ 0, & \text{otherwise.} \end{cases} \tag{3.111}$$

Note that large values of $\frac{\sigma^2_{v,m}(n)}{|\lambda_m(n)|^2}$ leads to low values of $\left(\sigma^{\text{o}}_{s,m}(n)\right)^2$, which means that less power is allocated to poor subchannels. The solution in expression (3.111) is the so-called *water-filling*. It essentially says that the power allocation among the subcarriers is implemented in such a way that $\left(\sigma^{\text{o}}_{s,m}(n)\right)^2 + \frac{\sigma^2_{v,m}(n)}{|\lambda_m(n)|^2}$ is a constant and that occasionally it will not be distributed any energy to some of the last subcarriers due to the power constraint. The name water-filling is motivated by the way it can be interpreted, namely, as communicating vessels which are filled with a homogeneous liquid that achieves the same level in all containers, regardless of their volume and/or shape.

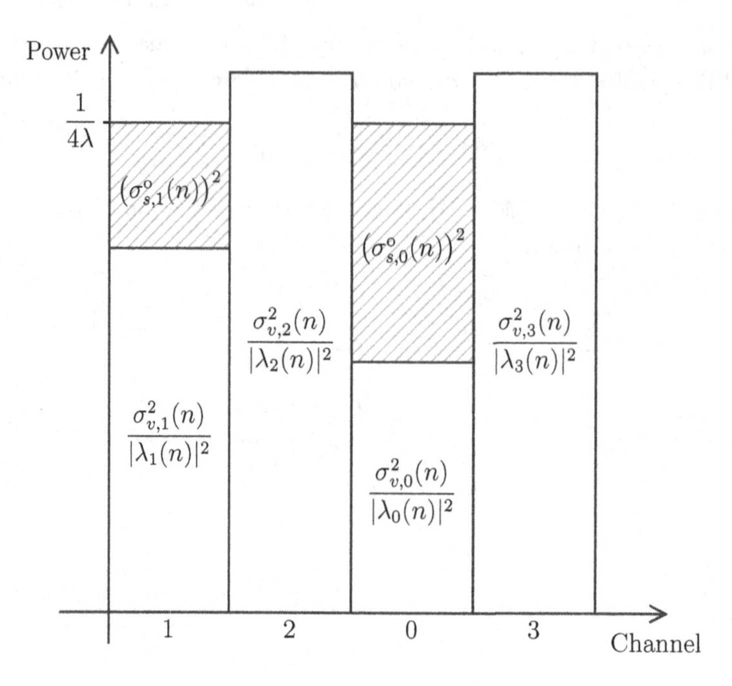

Figure 3.15: Discrete water-filling for four subchannels.

Figure 3.15 illustrates pictorially a water-filling solution for the case of four subchannels (i.e., $M = 4$) where the total energy distributed is $p_0(n) = \left(\sigma^{\text{o}}_{s,0}(n)\right)^2 + \left(\sigma^{\text{o}}_{s,1}(n)\right)^2$. The subchannels with indexes 3 and 4 did not receive any power (i.e., $M' = 2$) since the values of their corresponding factors $\frac{\sigma^2_{v,2}(n)}{|\lambda_2(n)|^2}$ and $\frac{\sigma^2_{v,3}(n)}{|\lambda_3(n)|^2}$ were both higher than $\frac{1}{4\lambda} = \frac{p_0(n)+\frac{\sigma^2_{v,0}(n)}{|\lambda_0(n)|^2}+\frac{\sigma^2_{v,1}(n)}{|\lambda_1(n)|^2}}{2}$. Example 3.2 is a toy example which help us verify the gains introduced by the water-filling approach.

Example 3.2 (Water-Filling) Given a time-invariant channel model with transfer function

$$H_n^{\text{d}}(z) = 1 - z^{-2}, \tag{3.112}$$

whose zeros are $e^{j\frac{2\pi}{8}0} = 1$ and $e^{j\frac{2\pi}{8}4} = -1$. Assume that the noise variance is given by $\sigma^2_{v,m}(n) = 0.08$ for all $m \in \{0, 1, \cdots, 7\}$ and $n \in \mathbb{Z}$. Considering eight subchannels, compute the channel capacity for each subchannel, $C_m(n)$, and average channel capacity, $C(n)$, before and after applying the water-filling approach. Assume two possible values of total transmission power, namely $p_0(n) = 0.04$ and $p_0(n) = 0.40$. In addition, determine the power distribution for each case.

Solution. In order to compute the initial SNR (i.e., the SNR before applying the water-filling optimization approach), one has to determine three quantities for each subchannel, namely $\sigma^2_{v,m}(n)$,

$\sigma_{s,m}^2(n)$, and $|\lambda_m(n)|^2$, as described in Equation (3.88). The noise variance is fixed at 0.08, while the average power of the symbols is $\sigma_{s,m}^2(n) = p_0(n)/8$, i.e., $\sigma_{s,m}^2(n) = 0.04/8 = 0.005$ or $\sigma_{s,m}^2(n) = 0.40/8 = 0.05$. As for $|\lambda_m(n)|^2$, one has (see Equation (3.67) or Equation (2.57) of Chapter 2)

$$|\lambda_m(n)|^2 = \left| \sqrt{M} \left[\mathbf{W}_M \begin{bmatrix} \mathbf{h}_n^d \\ \mathbf{0}_{5\times 1} \end{bmatrix} \right]_m \right|^2, \tag{3.113}$$

where $\mathbf{h}_n^d \triangleq [1 \ \ 0 \ \ -1]^T$.

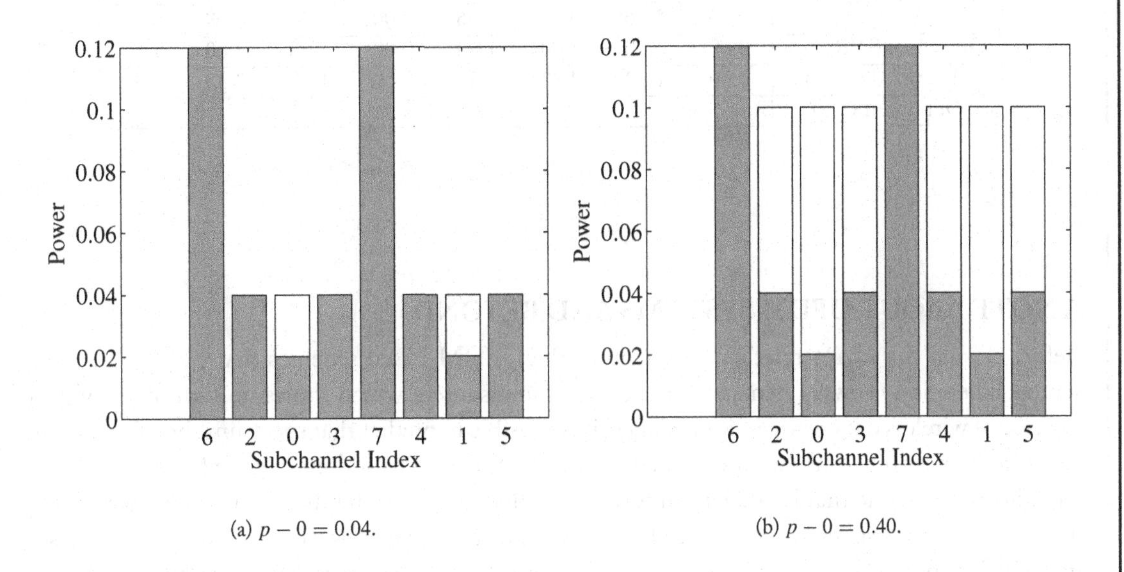

(a) $p - 0 = 0.04$. (b) $p - 0 = 0.40$.

Figure 3.16: Water-filling example.

The power distribution, the initial and optimized SNRs (the optimized SNR uses $\left(\sigma_{s,m}^{\text{o}}(n)\right)^2$ given in expression (3.111)), the initial and optimized capacity (see Equation (3.98)) for each subcarrier, and the average capacity (see Equation (3.99)) are summarized in the following table, considering both transmission powers:

Subchannel index	6	2	0	3	7	4	1	5
$\lvert\lambda_m(n)\rvert^2$	0.0	2.0	4.0	2.0	0.0	2.0	4.0	2.0
$\sigma^2_{v,m}(n)/\lvert\lambda_m(n)\rvert^2$	∞	0.04	0.02	0.04	∞	0.04	0.02	0.04
$p_0(n)$	0.04							
Initial $\sigma^2_{s,m}(n)$	0.005	0.005	0.005	0.005	0.005	0.005	0.005	0.005
Initial $\mathrm{SNR}_m(n)$	0.00	0.12	0.25	0.12	0.00	0.12	0.25	0.12
Initial $C_m(n)$	0.00	0.17	0.32	0.17	0.00	0.17	0.32	0.17
Initial $C(n)$	0.1654							
Optimized $\sigma^2_{s,m}(n)$	0.00	0.00	0.02	0.00	0.00	0.00	0.02	0.00
Optimized $\mathrm{SNR}_m(n)$	0.00	0.00	1.00	0.00	0.00	0.00	1.00	0.00
Optimized $C_m(n)$	0.00	0.00	1.00	0.00	0.00	0.00	1.00	0.00
Optimized $C(n)$	0.2500							
$p_0(n)$	0.40							
Initial $\sigma^2_{s,m}(n)$	0.05	0.05	0.05	0.05	0.05	0.05	0.05	0.05
Initial $\mathrm{SNR}_m(n)$	0.00	1.25	2.50	1.25	0.00	1.25	2.50	1.25
Initial $C_m(n)$	0.00	1.17	1.81	1.17	0.00	1.17	1.81	1.17
Initial $C(n)$	1.0368							
Optimized $\sigma^2_{s,m}(n)$	0.00	0.06	0.08	0.06	0.00	0.06	0.08	0.06
Optimized $\mathrm{SNR}_m(n)$	0.00	1.50	4.00	1.50	0.00	1.50	4.00	1.50
Optimized $C_m(n)$	0.00	1.32	2.32	1.32	0.00	1.32	2.32	1.32
Optimized $C(n)$	1.2414							

□

A NOTE ABOUT OFDM SYSTEMS AND BEYOND

Before closing this chapter, let us comment on the OFDM-based schemes that we have just presented. There is a growing demand for transmission resources which shows no sign of settling. In the case of wireless data services, for instance, it is possible to predict that spectrum shortage is a sure event in the near future. As we mentioned before, the first step to address this problem is to choose a modulation scheme that is efficient in terms of channel capacity in bits/transmission, particularly in broadband transmissions where the channel presents a frequency-selective model. In such cases, multicarrier communication systems is a very smart solution given that this modulation scheme is efficient for data transmission through channels with moderate and severe ISI.

The most widely used and simplest multicarrier system is the OFDM (comprising its variants presented in this chapter). Whenever the subchannels are narrow enough, it is possible to consider that each subchannel-frequency range is flat, avoiding the use of sophisticated equalizers. In applications where there is a return channel, it is also possible to exploit the SNR in different subchannels in order to select the subchannel-modulation order, so that subchannels with high SNR utilize high-order modulation, whereas low-order modulation should be employed in subchannels with moderate SNR. This loading strategy enables more effective usage of frequency-selective channels.

In OFDM, a serial data stream is divided into blocks where each block is appended with redundancy (guard interval) in order to avoid IBI. The special features of the OFDM system are the elimination of IBI, ISI, and its inherent low computational complexity. OFDM system utilizes IDFT at the transmitter and DFT at the receiver, enabling the use of computationally efficient FFT algorithms. The key strategy to avoid IBI in transmissions through frequency-selective channels is to include some redundancy at the transmitter, thus reducing the data rate.

The SC-FD scheme has also emerged as an alternative solution to overcome some drawbacks inherent to OFDM-based systems, such as PAPR and CFO. Also for some frequency-selective channels, the BER of an SC-FD system can be lower than for the OFDM, particularly if some subchannels have high attenuation. In this solution, the data stream is inserted with redundancy to avoid IBI and at the receiver the equalization is performed in the frequency domain, keeping the efficient equalization scheme inherited from OFDM.

A way to postpone the spectrum shortage is to increase the data throughput for a given bandwidth by utilizing some smart technology. Assuming we are employing block transmission it is worth discussing how to reduce the amount of redundancy in multicarrier system, while constraining the transceiver to employ fast algorithms. A possible solution is to employ reduced-redundancy transceivers utilizing DFTs and diagonal matrices. The reduced-redundancy transceiver allows for higher data throughput than OFDM in a number of practical situations. A computationally efficient solution for reduced-redundancy transceiver is presented in Chapter 4.

3.5 CONCLUDING REMARKS

In this chapter we addressed distinct aspects of OFDM systems, for example, continuous-time (analog) OFDM, discrete-time OFDM, coded OFDM (C-OFDM), and DMT. We derived the discrete-time OFDM as the sampled version of the analog OFDM and discussed the issue of ensuring orthogonality among subcarriers in frequency-selective channels. The C-OFDM was introduced as a discrete-time OFDM in which we include some channel encoding elements (inner and outer coders, interleavers and/or scramblers) in order to protect the information to be transmitted. In addition, the DMT can be seen as a discrete-time OFDM system in which the transmitter has channel state information, and, therefore, it is capable of performing optimal power allocation, thus increasing the channel capacity given a power constraint.

All these systems require the use of a guard period whose time extension is greater or equal to the channel memory (or time delay spread, for the analog OFDM). Therefore, the guard period τ increases with the channel memory τ_{mem}. As a consequence, for an OFDM symbol of a given duration T, the useful symbol time $T - \tau$ decreases, thus reducing the system capacity. In other words, when the durations of the OFDM symbol and the channel impulse response are of the same order, the throughput of OFDM systems decreases. In order to address this issue, the next chapter describes systems whose computational complexity are comparable to the OFDM complexity, with the advantage of employing a reduced number of redundant elements, that is, their guard time is shorter.

CHAPTER 4

Memoryless LTI Transceivers with Reduced Redundancy

4.1 INTRODUCTION

Multicarrier-modulation methods are of paramount importance to many data-transmission systems whose channels induce severe or moderate intersymbol interference (ISI). As previously discussed, the key idea behind the success of multicarrier techniques is the partition of the physical channel into (ideally) non-interfering narrowband subchannels. If the bandwidths associated with these subchannels are narrow enough, then the related channel-frequency response of each subchannel appears to be flat, thus avoiding the use of sophisticated equalizers. In addition, the subchannel division allows, whenever possible, the exploitation of signal-to-noise ratios (SNRs) in the different subchannels by managing the data load in each individual subchannel.

The orthogonal frequency-division multiplexing (OFDM) described in Chapter 3 is the most popular multicarrier-modulation technique. OFDM-based systems feature lots of good properties regarding their performance and implementation simplicity. However, the OFDM has some drawbacks, such as high peak-to-average power ratio (PAPR), high sensitivity to carrier-frequency offset (CFO), and (possibly) significant loss in spectral efficiency due to the redundancy insertion required to eliminate the interblock interference (IBI). The single-carrier with frequency-domain equalization (SC-FD) is an efficient transmission technique which reduces both PAPR and CFO as compared to the OFDM system. These advantages are attained without changing drastically the overall complexity of the transceiver, as shown in Chapter 3.

Alternatively, one could consider the general transmultiplexer (TMUX) framework described in Chapter 2 in order to conceive new multicarrier-modulation techniques which are able to circumvent some of the OFDM limitations. For example, intuition might suggest that we can reduce intercarrier interference (ICI) caused by the loss of orthogonality between the subcarriers as long as we are able to design TMUXes containing a large number of subchannels with sharp transitions. Therefore, it would be possible to achieve better solutions (in terms of bit-error rate, for example) for filter-bank transceivers to be employed in both future generations of wireless systems and current developments of local/broadband wireless networks. This strategy, however, remains to be proved as a viable solution in practice. In fact, one should always take into account the fundamental trade-off between performance gains and cost effectiveness from a practical perspective. The computational complexity[1] is among the factors that directly affects the cost effectiveness of new advances in com-

[1]Number of arithmetical operations employed in the related processing.

munications. TMUXes with sharp transitions must necessarily have long memory, which might hinder their use in practice. This explains why memoryless linear time-invariant (LTI) TMUXes are still preferred in many practical applications.

The price paid for using memoryless LTI TMUXes is that some redundancy must be necessarily included at the transmitter end in order to allow the elimination of the ISI induced by the channel. Redundancy plays a central role in communications systems. Channel-coding schemes are good examples of how to apply redundancy in order to achieve reliable transmissions. In addition, redundancy is also employed in many block-based transceivers in practice, such as OFDM and SC-FD systems, in order to eliminate the inherent IBI and to yield simple equalizer structures.

Regarding the spectral-resource usage, the amount of redundancy employed in both OFDM and SC-FD systems depends on the delay spread of the channel, implying that both transceivers waste the same bandwidth on redundant data. Nevertheless, there are many ways to increase the spectral efficiency of communications systems, such as by decreasing the overall symbol-error probability in the physical layer, so that less redundancy needs to be inserted in upper-layers by means of channel coding. In general, this approach increases the costs in the physical layer, since it leads to more computationally complex transceivers, hindering its implementation in some practical applications.

Other means to improve spectral efficiency are, therefore, highly desirable. Reducing the amount of transmitted redundancy inserted in the physical layer is *a possible solution*.[2] Although reliability and simplicity are rather important in practical applications, the amount of redundancy should be reduced to potentially increase the spectral efficiency. In the context of fixed and memoryless TMUXes, it is possible to show that the minimum redundancy required to eliminate IBI and still allow the design of zero-forcing (ZF) solutions is only half the amount of redundancy used by standard OFDM and SC-FD systems.

In this context, an important question arises: can we design memoryless LTI transceivers with reduced redundancy whose computational complexity is comparable to OFDM and SC-FD systems? That is, these transceivers should be amenable to superfast implementations in order to keep their computational complexities competitive with practical OFDM-based systems. If this is possible, the resulting transceivers would probably allow higher data throughputs in broadband channels.

This chapter describes how transceivers with reduced redundancy (Section 4.2) can be implemented employing superfast algorithms based on the concepts of structured matrix representations (Section 4.3). To achieve this objective, we describe some mathematical decompositions of a special class of structured matrix: the so-called Bezoutian matrices (Section 4.4). The resulting structures of multicarrier and single-carrier reduced-redundancy systems are then presented and analyzed through some examples (Section 4.5).

[2]The bottom line here is that many distinct and interesting ways of designing multicarrier systems are available. We will focus on a particular type of solution (reduced-redundancy system) that allows us to present to the reader a set of tools related to structured matrix representations, which can be eventually employed in many other contexts.

4.2 REDUCED-REDUNDANCY SYSTEMS: THE ZP-ZJ MODEL REVISITED

As briefly described in Chapter 2, Subsection 2.4.5, Lin and Phoong proposed a family of memoryless discrete multitone transceivers which employ a reduced amount of redundancy to eliminate both IBI and ISI [44]. A useful particular type of reduced-redundancy system is the so-called *zero-padded zero-jammed (ZP-ZJ) system*. Let us revisit now the ZP-ZJ model in a more detailed manner.

Consider a complex-valued finite impulse-response (FIR) channel model $h(l)$, where the channel-tap index l can assume any value within the set $\mathcal{L} \triangleq \{0, 1, \cdots, L\}$. The integer number L is the so-called *channel order*, which in turn is associated with the *delay spread* of the channel [80]: a measure of how much a signal is spread over the time when passing through the channel.

In ZP-ZJ systems, the transmitter is responsible for linearly processing the input vector $\mathbf{s}(n) \in \mathcal{C}^{M\times 1}$, where the integer number n is the time index, the natural number M denotes the amount of symbols to be transmitted, and $\mathcal{C} \subset \mathbb{C}$ is a given digital constellation. Such a processing is implemented through the multiplication by the matrix

$$\mathbf{F} \triangleq \begin{bmatrix} \overline{\mathbf{F}} \\ \mathbf{0}_{K\times M} \end{bmatrix} \in \mathbb{C}^{N\times M}, \tag{4.1}$$

where $N \in \mathbb{N}$ is the total length of the transmitted message, whereas $\overline{\mathbf{F}} \in \mathbb{C}^{M\times M}$ is the *actual transmitter matrix*, or just the *transmitter matrix*, which is in charge of preparing the vector for the transmission through the FIR channel. The matrix $\mathbf{0}_{K\times M}$ is responsible for adding the redundant zero elements (*zero-padding process*). The amount of redundancy inserted is the natural number

$$K \triangleq N - M. \tag{4.2}$$

If one assumes that the channel order L is less than the total length of the transmitted message N, then the received block $\mathbf{y}(n) \in \mathbb{C}^{N\times 1}$ will suffer from ISI and IBI effects according to the channel model described in Chapter 2, Equation (2.39), which can be rewritten as

$$\mathbf{y}(n) = \mathbf{H}_{\text{ISI}}\mathbf{F}\mathbf{s}(n) + \mathbf{H}_{\text{IBI}}\mathbf{F}\mathbf{s}(n-1) + \mathbf{v}(n), \tag{4.3}$$

where $\mathbf{v}(n) \in \mathbb{C}^{N\times 1}$ accounts for an additive noise at the receiver front-end, as depicted, for example, in Figure 2.9 of Chapter 2.

In order to generate an estimate $\hat{\mathbf{s}}(n) \in \mathbb{C}^{M\times 1}$ of the transmitted data vector,[3] the receiver applies another linear transformation to the incoming signal vector through the matrix

$$\mathbf{G} \triangleq \begin{bmatrix} \mathbf{0}_{M\times(L-K)} & \overline{\mathbf{G}} \end{bmatrix} \in \mathbb{C}^{M\times N}, \tag{4.4}$$

where $\overline{\mathbf{G}} \in \mathbb{C}^{M\times(N+K-L)}$ is the *actual receiver matrix*, or just the *receiver matrix*. The matrix $\mathbf{0}_{M\times(L-K)}$ is responsible for removing the remaining IBI (*zero-jamming process*) that was not completely eliminated by the transmitted redundancy.

[3]Observe that the estimated elements do not have to pertain to the digital constellation $\mathcal{C}$. Only after a hard-decision-detection process the estimated elements are mapped into constellation symbols belonging to $\mathcal{C}$.

Now, by using the definitions of $\mathbf{H}_{\text{ISI}}$ and $\mathbf{H}_{\text{IBI}}$ given in Equation (2.38) of Chapter 2, the previous ZP-ZJ model description yields the following estimate:

$$\begin{aligned}
\hat{\mathbf{s}}(n) &\triangleq \mathbf{G}\mathbf{y}(n) \\
&= \mathbf{G}\mathbf{H}_{\text{ISI}}\mathbf{F}\mathbf{s}(n) + \underbrace{\mathbf{G}\mathbf{H}_{\text{IBI}}\mathbf{F}}_{=\mathbf{0}_{M\times M}}\mathbf{s}(n-1) + \underbrace{\mathbf{G}\mathbf{v}(n)}_{\triangleq\overline{\mathbf{v}}(n)} \\
&= \mathbf{G}\mathbf{H}_{\text{ISI}}\mathbf{F}\mathbf{s}(n) + \overline{\mathbf{v}}(n) \\
&= \begin{bmatrix}\mathbf{0}_{M\times(L-K)} & \overline{\mathbf{G}}\end{bmatrix}\mathbf{H}_{\text{ISI}}\begin{bmatrix}\overline{\mathbf{F}} \\ \mathbf{0}_{K\times M}\end{bmatrix}\mathbf{s}(n) + \overline{\mathbf{v}}(n) \\
&= \underbrace{\overline{\mathbf{G}}\,\overline{\mathbf{H}}\,\overline{\mathbf{F}}}_{\triangleq\overline{\mathbf{T}}}\mathbf{s}(n) + \overline{\mathbf{v}}(n) \\
&= \overline{\mathbf{T}}\mathbf{s}(n) + \overline{\mathbf{v}}(n),
\end{aligned} \tag{4.5}$$

where the equivalent *channel matrix* $\overline{\mathbf{H}}$ is defined as

$$\overline{\mathbf{H}} \triangleq \begin{bmatrix}
h(L-K) & \cdots & h(0) & 0 & 0 & \cdots & 0 \\
\vdots & \ddots & & & & & \vdots \\
h(K) & \ddots & & & & & 0 \\
\vdots & \ddots & & & \ddots & & h(0) \\
h(L) & & & & & & \vdots \\
0 & & & & \ddots & & h(L-K) \\
\vdots & & & & & & \vdots \\
0 & \cdots & 0 & 0 & h(L) & \cdots & h(K)
\end{bmatrix} \in \mathbb{C}^{(N+K-L)\times M}. \tag{4.6}$$

Equation (4.5) shows that the estimated vector $\hat{\mathbf{s}}(n)$ does not depend on the transmitted vector $\mathbf{s}(n-1)$, which means that the overall transmission/reception process is performed on a block-by-block basis. We can therefore simplify the notation by omitting the time-dependency of the related variables. Thus, Equation (4.5) can be rewritten as

$$\hat{\mathbf{s}} = \overline{\mathbf{T}}\mathbf{s} + \overline{\mathbf{v}}. \tag{4.7}$$

Figure 4.1 illustrates the ZP-ZJ model omitting the time-dependency of the related variables. Note that the choices for the dimensions of the all-zero matrices which implement the zero-padding zero-jamming processes make sense. Indeed, by remembering the definition of matrix $\mathbf{H}_{\text{IBI}}$ in Equation (2.38), we can verify that only the first L elements of a received data block are affected by the last L elements of the previous transmitted data block. This interblock interference occurs due to the memory of the channel. Hence, the IBI effect can be eliminated by discarding the first L out of N elements from all received data blocks. However, if one desires to transmit $N = M + K$ elements,

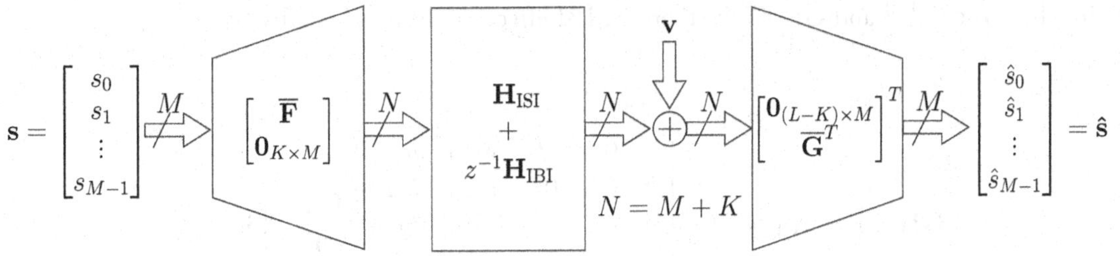

Figure 4.1: ZP-ZJ transceiver model.

with K *redundant zeros* included at the end of each transmitted block, then instead of discarding the first L elements from the received data block, one can discard only the first $L-K$ elements, since the immediately following $L-(L-K)=K$ elements are "affected" by the zero-valued transmitted elements appended at the end of the previous transmitted block (see Example 4.1 below).

In general, K can assume any non-negative integer value in order to eliminate the IBI effect. In fact, if $K \geq L$, then we must adjust the notation employed to define matrix $\mathbf{G}$ in Equation (4.4), since $L-K \leq 0$. In such a case, we could redefine $\mathbf{G}$ without zero jamming, i.e., $\mathbf{G}=\overline{\mathbf{G}} \in \mathbb{C}^{M\times N}$. By doing so, we see that ZP-ZJ systems encompass general ZP-OFDM and ZP-SC-FD systems which transmit $K \geq L$ redundant zeros.

Example 4.1 (Zero-Padding Zero-Jamming Process) Consider an FIR channel with transfer function

$$H(z)=h(0)+h(1)z^{-1}+h(2)z^{-2}+h(3)z^{-3}. \tag{4.8}$$

In this case, we have $L=3$. Now, assume that $M=4$ and $K=1$, implying that $N=M+K=5$. Thus, we have (see Equation (2.38))

$$\mathbf{H}(z)=\underbrace{\begin{bmatrix} h(0) & 0 & 0 & 0 & 0 \\ h(1) & h(0) & 0 & 0 & 0 \\ h(2) & h(1) & h(0) & 0 & 0 \\ h(3) & h(2) & h(1) & h(0) & 0 \\ 0 & h(3) & h(2) & h(1) & h(0) \end{bmatrix}}_{=\mathbf{H}_{\text{ISI}}} + z^{-1} \underbrace{\begin{bmatrix} 0 & 0 & h(3) & h(2) & h(1) \\ 0 & 0 & 0 & h(3) & h(2) \\ 0 & 0 & 0 & 0 & h(3) \\ 0 & 0 & 0 & 0 & 0 \\ 0 & 0 & 0 & 0 & 0 \end{bmatrix}}_{=\mathbf{H}_{\text{IBI}}}. \tag{4.9}$$

Observe that, if we use the following matrices (see Equations (4.1) and (4.4))

$$\mathbf{F}=\begin{bmatrix} \mathbf{I}_4 \\ \mathbf{0}_{1\times 4} \end{bmatrix} \overline{\mathbf{F}}, \tag{4.10}$$

$$\mathbf{G}=\begin{bmatrix} \mathbf{0}_{4\times 2} & \overline{\mathbf{G}} \end{bmatrix}, \tag{4.11}$$

in which $\overline{\mathbf{F}} \in \mathbb{C}^{4\times 4}$ and $\overline{\mathbf{G}} \in \mathbb{C}^{4\times 3}$, then the IBI effect can be eliminated since

$$
\begin{aligned}
\mathbf{G}\mathbf{H}_{\text{IBI}}\mathbf{F} &= \begin{bmatrix} \mathbf{0}_{4\times 2} & \overline{\mathbf{G}} \end{bmatrix} \begin{bmatrix} 0 & 0 & h(3) & h(2) & h(1) \\ 0 & 0 & 0 & h(3) & h(2) \\ 0 & 0 & 0 & 0 & h(3) \\ 0 & 0 & 0 & 0 & 0 \\ 0 & 0 & 0 & 0 & 0 \end{bmatrix} \begin{bmatrix} \mathbf{I}_4 \\ \mathbf{0}_{1\times 4} \end{bmatrix} \overline{\mathbf{F}} \\
&= \begin{bmatrix} \mathbf{0}_{4\times 2} & \overline{\mathbf{G}} \end{bmatrix} \left(\begin{bmatrix} 0 & 0 & h(3) & h(2) \\ 0 & 0 & 0 & h(3) \\ 0 & 0 & 0 & 0 \\ 0 & 0 & 0 & 0 \\ 0 & 0 & 0 & 0 \end{bmatrix} \mathbf{I}_4 + \underbrace{\begin{bmatrix} h(1) \\ h(2) \\ h(3) \\ 0 \\ 0 \end{bmatrix} \mathbf{0}_{1\times 4}}_{=\mathbf{0}_{5\times 4}} \right) \overline{\mathbf{F}} \\
&= \begin{bmatrix} \mathbf{0}_{4\times 2} & \overline{\mathbf{G}} \end{bmatrix} \begin{bmatrix} 0 & 0 & h(3) & h(2) \\ 0 & 0 & 0 & h(3) \\ 0 & 0 & 0 & 0 \\ 0 & 0 & 0 & 0 \\ 0 & 0 & 0 & 0 \end{bmatrix} \overline{\mathbf{F}} \\
&= \left(\underbrace{\mathbf{0}_{4\times 2} \begin{bmatrix} 0 & 0 & h(3) & h(2) \\ 0 & 0 & 0 & h(3) \end{bmatrix}}_{=\mathbf{0}_{4\times 4}} + \underbrace{\overline{\mathbf{G}}\, \mathbf{0}_{3\times 4}}_{=\mathbf{0}_{4\times 4}} \right) \overline{\mathbf{F}} \\
&= \mathbf{0}_{4\times 4}.
\end{aligned} \tag{4.12}
$$

Moreover, it is also possible to verify that the structure of the equivalent channel matrix $\overline{\mathbf{H}}$ described in Equation (4.6) holds in this example. Indeed, we have

$$
\mathbf{G}\mathbf{H}_{\text{ISI}}\mathbf{F} = \begin{bmatrix} \mathbf{0}_{4\times 2} & \overline{\mathbf{G}} \end{bmatrix} \begin{bmatrix} h(0) & 0 & 0 & 0 & 0 \\ h(1) & h(0) & 0 & 0 & 0 \\ h(2) & h(1) & h(0) & 0 & 0 \\ h(3) & h(2) & h(1) & h(0) & 0 \\ 0 & h(3) & h(2) & h(1) & h(0) \end{bmatrix} \begin{bmatrix} \mathbf{I}_4 \\ \mathbf{0}_{1\times 4} \end{bmatrix} \overline{\mathbf{F}}
$$

$$
\begin{aligned}
&= \begin{bmatrix} \mathbf{0}_{4\times 2} & \overline{\mathbf{G}} \end{bmatrix} \left(\begin{bmatrix} h(0) & 0 & 0 & 0 \\ h(1) & h(0) & 0 & 0 \\ h(2) & h(1) & h(0) & 0 \\ h(3) & h(2) & h(1) & h(0) \\ 0 & h(3) & h(2) & h(1) \end{bmatrix} \mathbf{I}_4 + \underbrace{\begin{bmatrix} 0 \\ 0 \\ 0 \\ 0 \\ h(0) \end{bmatrix} \mathbf{0}_{1\times 4}}_{=\mathbf{0}_{5\times 4}} \right) \overline{\mathbf{F}} \\
&= \begin{bmatrix} \mathbf{0}_{4\times 2} & \overline{\mathbf{G}} \end{bmatrix} \begin{bmatrix} h(0) & 0 & 0 & 0 \\ h(1) & h(0) & 0 & 0 \\ h(2) & h(1) & h(0) & 0 \\ h(3) & h(2) & h(1) & h(0) \\ 0 & h(3) & h(2) & h(1) \end{bmatrix} \overline{\mathbf{F}} \\
&= \left(\underbrace{\mathbf{0}_{4\times 2} \begin{bmatrix} h(0) & 0 & 0 & 0 \\ h(1) & h(0) & 0 & 0 \end{bmatrix}}_{=\mathbf{0}_{4\times 4}} + \overline{\mathbf{G}} \underbrace{\begin{bmatrix} h(2) & h(1) & h(0) & 0 \\ h(3) & h(2) & h(1) & h(0) \\ 0 & h(3) & h(2) & h(1) \end{bmatrix}}_{=\overline{\mathbf{H}} \in \mathbb{C}^{3\times 4}} \right) \overline{\mathbf{F}} \\
&= \overline{\mathbf{G}}\,\overline{\mathbf{H}}\,\overline{\mathbf{F}}.
\end{aligned} \tag{4.13}
$$

□

As we have seen in Example 4.1, the redundancy is padded at the transmitter and jammed at the receiver end in such a way that the IBI effect is completely eliminated. Nevertheless, the amount of redundancy should be such that the ISI effect might also be eliminated, i.e., the amount of redundant zeros should not be too small, otherwise it would prevent us from finding ZF solutions (or, in other words, ISI elimination). Indeed, the definition of the $M \times M$ *equivalent transfer matrix* $\overline{\mathbf{T}}$ in Equation (4.5) makes clear that a zero-forcing solution is achievable if, and only if, the matrix $\overline{\mathbf{T}} = \overline{\mathbf{G}}\,\overline{\mathbf{H}}\,\overline{\mathbf{F}}$ is full-rank (i.e., $\text{rank}\{\overline{\mathbf{T}}\} = M$). Since the dimensions of matrices $\overline{\mathbf{G}}$, $\overline{\mathbf{H}}$, and $\overline{\mathbf{F}}$ are $M \times (M + 2K - L)$, $(M + 2K - L) \times M$, and $M \times M$, respectively, and since the rank of the product of matrices is not larger than the rank of each individual matrix, then we must necessarily have

$$
M \leq \min \{M, M + 2K - L\}, \tag{4.14}
$$

implying $M \leq M + 2K - L \Leftrightarrow 0 \leq 2K - L$. This result means that

$$
K \geq \left\lceil \frac{L}{2} \right\rceil, \tag{4.15}
$$

where the notation $\lceil x \rceil$ denotes the ceiling operator, which gives us the unique integer number within the interval $[x, x+1) \subset \mathbb{R}$, for any real number x. Notice that the ZF solution is only achieved if matrices $\overline{\mathbf{G}}$, $\overline{\mathbf{H}}$, and $\overline{\mathbf{F}}$ have full-rank.[4]

Consider that we do insert at least $\lceil L/2 \rceil$ zeros before the transmission and assume that channel-state information (CSI) is available at the receiver end, i.e., $h(l)$ is known for all $l \in \mathcal{L}$. In addition, assume that the transmitter employs a predefined *channel-independent unitary precoder* $\overline{\mathbf{F}}$,[5] so that the task of defining the rectangular matrix $\overline{\mathbf{G}}$ is left to the designer. In this context, the usual objective is to minimize either the ISI or the mean square error (MSE) at the receiver. Such goals can be achieved by the ZF and the minimum MSE (MMSE) receivers, respectively. These solutions are

$$\begin{aligned}
\overline{\mathbf{G}}_{\text{ZF}} &\triangleq (\overline{\mathbf{H}}\,\overline{\mathbf{F}})^{\dagger} \\
&= [(\overline{\mathbf{H}}\,\overline{\mathbf{F}})^H (\overline{\mathbf{H}}\,\overline{\mathbf{F}})]^{-1} (\overline{\mathbf{H}}\,\overline{\mathbf{F}})^H \\
&= [\overline{\mathbf{F}}^H \overline{\mathbf{H}}^H \overline{\mathbf{H}}\,\overline{\mathbf{F}}]^{-1} \overline{\mathbf{F}}^H \overline{\mathbf{H}}^H \\
&= \overline{\mathbf{F}}^{-1} (\overline{\mathbf{H}}^H \overline{\mathbf{H}})^{-1} \left(\overline{\mathbf{F}}^H\right)^{-1} \overline{\mathbf{F}}^H \overline{\mathbf{H}}^H \\
&= \overline{\mathbf{F}}^{-1} \underbrace{(\overline{\mathbf{H}}^H \overline{\mathbf{H}})^{-1} \overline{\mathbf{H}}^H}_{\triangleq \overline{\mathbf{H}}^{\dagger}} \\
&= \overline{\mathbf{F}}^H \overline{\mathbf{H}}^{\dagger},
\end{aligned} \tag{4.16}$$

$$\begin{aligned}
\overline{\mathbf{G}}_{\text{MMSE}} &\triangleq \left[(\overline{\mathbf{H}}\,\overline{\mathbf{F}})^H (\overline{\mathbf{H}}\,\overline{\mathbf{F}}) + \frac{\sigma_v^2}{\sigma_s^2} \mathbf{I}_M\right]^{-1} (\overline{\mathbf{H}}\,\overline{\mathbf{F}})^H \\
&= \left(\overline{\mathbf{F}}^H \overline{\mathbf{H}}^H \overline{\mathbf{H}}\,\overline{\mathbf{F}} + \frac{\sigma_v^2}{\sigma_s^2} \overline{\mathbf{F}}^H \overline{\mathbf{F}}\right)^{-1} \overline{\mathbf{F}}^H \overline{\mathbf{H}}^H \\
&= \overline{\mathbf{F}}^{-1} \left(\overline{\mathbf{H}}^H \overline{\mathbf{H}} + \frac{\sigma_v^2}{\sigma_s^2} \mathbf{I}_M\right)^{-1} \left(\overline{\mathbf{F}}^H\right)^{-1} \overline{\mathbf{F}}^H \overline{\mathbf{H}}^H \\
&= \overline{\mathbf{F}}^H \left(\overline{\mathbf{H}}^H \overline{\mathbf{H}} + \frac{\sigma_v^2}{\sigma_s^2} \mathbf{I}_M\right)^{-1} \overline{\mathbf{H}}^H,
\end{aligned} \tag{4.17}$$

where $(\cdot)^{\dagger}$ denotes the Moore-Penrose pseudo-inverse of $(\cdot)$. The optimal MMSE solution is derived following the same procedure as in Subsection 2.4.1. In addition, the vectors $\mathbf{s}$ and $\overline{\mathbf{v}}$ are considered uncorrelated wide-sense stationary (WSS) random sequences with zero-mean,[6] yielding $E[\mathbf{s}\overline{\mathbf{v}}]^H = E[\mathbf{s}]E[\overline{\mathbf{v}}]^H = \mathbf{0}_{M\times M} = E[\overline{\mathbf{v}}]E[\mathbf{s}]^H = E[\overline{\mathbf{v}}\mathbf{s}^H]$. We also assume that $E[\mathbf{s}\mathbf{s}^H] = \sigma_s^2 \mathbf{I}_M$ and $E[\overline{\mathbf{v}}\,\overline{\mathbf{v}}^H] = \sigma_v^2 \mathbf{I}_M$, in which $\sigma_v^2, \sigma_s^2 \in \mathbb{R}_+$ denote the variances of the random sequences (they do not depend on time because the random sequences are WSS). The given expression for $\overline{\mathbf{G}}_{\text{ZF}}$ is

[4] In Example 4.1, we see that, even though IBI is eliminated, it is not possible to achieve ISI elimination since matrix $\overline{\mathbf{G}} \in \mathbb{C}^{4\times 3}$ and $\overline{\mathbf{H}} \in \mathbb{C}^{3\times 4}$ have ranks at most 3, instead of $4 = M$. This happens since we adopted $K = 1 < \left\lceil \frac{L}{2} \right\rceil = 2$, thus violating the minimum required redundancy.

[5] The unitary property means $\overline{\mathbf{F}}^{-1} = \overline{\mathbf{F}}^H$.

[6] Remember that the time index was omitted for the sake of conciseness.

valid in the case $\overline{\mathbf{H}}$ has full-rank. As a result, we are assuming that the FIR channel model leads to a rank-M matrix $\overline{\mathbf{H}}$.

The ZF design does not require any estimate of the environment-noise variance and it is mainly discussed for theoretical purposes. This lack of information about the noise variance leads to poor performance when the SNR is low at the receiver end. On the other hand, the practical MMSE design is computationally more demanding than the ZF solution, but leads to more effective equalization.

The next step is to discuss how complex is performing both the *equalization* of a received data block and the *design* of the related receiver. The equalization consists of processing the received vector after the zero-jamming process through its multiplication by matrix $\overline{\mathbf{G}}$. This operation entails $\mathcal{O}(M^2)$ complex-valued arithmetical operations for general unstructured matrices. The equalization process requires the knowledge or the estimation of $\overline{\mathbf{H}}$ and possibly its regularized pseudo-inverse. This estimation is made during the receiver design. As a result, the computational complexity of the receiver design is $\mathcal{O}(M^3)$ complex-valued arithmetical operations for general unstructured matrices. The cyclic-prefix OFDM (CP-OFDM) and the CP-SC-FD transceivers have the advantage of performing the equalization as well as the receiver design employing only $\mathcal{O}(M \log_2 M)$ complex-valued operations due to their structural simplicity.

The use of discrete Fourier transform (DFT) and inverse DFT (IDFT) in order to decouple the estimation of the symbols at the receiver end are paramount to the success of CP-OFDM-based systems. Unfortunately, we cannot decouple so easily the estimation of the symbols in a ZP-ZJ system with reduced redundancy. Indeed, such decoupling process would require the computation of singular-value decompositions (SVDs), hindering its implementation in several practical problems.

Despite this potential drawback, we shall describe some tools that allow one to implement low complexity ZP-ZJ systems with reduced redundancy. As a motivating example, let us consider how a zero-forcing CP-SC-FD system is implemented. The insertion of the cyclic prefix turns the linear convolution into a circular convolution between the transmitted data symbols and the channel-impulse response. Using the vector notation for a noiseless channel, we can write $\mathbf{y} \triangleq \mathbf{H}_c\mathbf{s}$, where $\mathbf{H}_c$ is a circulant matrix which contains the channel coefficients. From linear algebra, we know that all circulant matrices may be diagonalized by using the same set of orthonormal eigenvectors. These eigenvectors are the columns of the unitary DFT matrix, $\mathbf{W}$. In addition, the eigenvalues of circulant channel matrices are easily computed by means of the DFT of the first column of the circulant matrix (see the discussions in Subsection 2.4.1). Thus, we have $\mathbf{y} = \mathbf{H}_c\mathbf{s} = \mathbf{W}^H\mathbf{\Lambda}\mathbf{W}\mathbf{s}$, which implies that we can recover $\mathbf{s}$ by performing $\mathbf{s} = \mathbf{H}_c^{-1}\mathbf{y} = \mathbf{W}^H\mathbf{\Lambda}^{-1}\mathbf{W}\mathbf{y}$, considering that $\mathbf{\Lambda}^{-1}$ is computable, i.e., all eigenvalues of $\mathbf{H}_c$ are non-zero. Hence, the ZF-SC-FD system that employs cyclic prefix *decomposes the inverse of the equivalent channel matrix using DFT and diagonal matrices.* In fact, this decomposition is quite special since it is a diagonalization of the inverse of the equivalent channel matrix.

Our aim is to follow a similar approach, that is, to look for an efficient decomposition for the "inverse" of the equivalent channel matrix associated with ZP-ZJ systems with reduced redundancy.

In such systems, the equivalent channel matrix $\overline{\mathbf{H}}$ is no longer circulant; rather, it is an $(M + 2K - L) \times M$ *Toeplitz matrix*, as described in Equation (4.6). Nevertheless, we could take into account the Toeplitz structure in order to decompose the generalized inverse of $\overline{\mathbf{H}}$, maybe using only DFT and diagonal matrices. Such an approach employs the same basic ideas present in CP-OFDM-based systems, *except* for two main features present only in OFDM-based systems: (i) the inverse of the equivalent channel matrix has exactly the same structure as the equivalent channel matrix itself (circulant structure); and (ii) the efficient decomposition of the inverse of the equivalent channel matrix corresponds to its diagonalization.

4.3 STRUCTURED MATRIX REPRESENTATIONS

A matrix is considered *structured* if its entries follow some predefined pattern. The origin of this pattern can be elementary mathematical relations among the matrix coefficients or simply the way certain coefficients appear repeatedly as entries of the given matrix. Some of these patterns happen naturally in the matrices describing the behavior of many practical systems (e.g., linear and circular discrete-time convolutions can be modeled by structured matrices) or can be easily induced by applying simple matrix transformations.

In any case, a structured matrix can be described by using a reduced set of distinct parameters, that is, the number of parameters required to describe the matrix is much smaller than its number of entries. There are several examples of structured matrices which are usually found in signal-processing and communication applications, such as: diagonal, circulant, pseudo-circulant, Toeplitz, among others.

Such structural patterns may bring about efficient means for exploiting features of the related problems. Besides, computations involving structured matrices can be further simplified by taking into account these structural patterns. Consider, for instance, the sum of two $M_2 \times M_1$ Toeplitz matrices. If one ignores the structural patterns present in such matrices, then this operation will require $M_2 M_1$ additions, since there are $M_2 M_1$ entries in each matrix. However, these matrices are completely defined by up to $M_2 + M_1 - 1$ elements, since the first row (M_1 elements) along with the first column (M_2 elements, in which the first element pertains to the first row as well) are enough to define a given Toeplitz matrix. This way, it would be quite reasonable to expect that matrix operations may be performed faster by using a reduced amount of parameters. Indeed, if one considers the structure of the matrices, then this operation will require only $M_2 + M_1 - 1$ additions corresponding to the sum of the first row and the first column of each matrix. The resulting Toeplitz matrix can be built by rearranging the elements of the resulting vectors accordingly.

As previously mentioned, the widespread use of OFDM and SC-FD transceivers relies on their key feature of transforming the original description of the Toeplitz channel-convolution matrix into a circulant matrix, for the case where the channel is linear and time-invariant.[7] Since a circulant matrix has eigenvectors comprising the columns of the unitary DFT matrix, the diagonalization

[7]The time-invariance assumption only needs to hold during the transmission of one data block.

process is quite efficient due to the existence of fast Fourier transform (FFT) algorithms, turning the equalization process very simple.

The effective channel matrix associated with ZP-ZJ systems is a rectangular Toeplitz matrix (see Equation (4.6)). It is therefore natural to expect that linear equalizers, such as linear MMSE or ZF equalizers, can take advantage of the structure of this channel matrix like OFDM-based systems do. In this context, three questions arise: (i) How to recognize a structured matrix by using analytical tools? (ii) How to represent the linear optimal solutions (either MMSE or ZF) by employing such analytical tools? and (iii) How to effectively take advantage of such representations?

This section answers those questions in the context of rectangular structured matrices. A useful tool for exploiting the structure of a matrix is the *displacement approach* proposed by Kailath et al. [34]. This section addresses the representation of structured matrices using the displacement structure. Indeed, we apply the concept of displacement to some structured matrices that appear in reduced-redundancy transceivers. An important result is the decomposition of *Bezoutian matrices*, because they play a key role in the design of transceivers using reduced redundancy.

4.3.1 DISPLACEMENT-RANK APPROACH

There are several ways to measure how much a matrix is structured. One of the most widely used is the *displacement operator*. The structural property of a matrix can be determined by analyzing the matrix resulting from the application of the displacement operation on the original matrix. Such analysis can be as simple as verifying the rank of the resulting matrix after the application of the displacement operator.

While studying techniques to solve Toeplitz-like problems, Kailath et al. [34] proposed a procedure to measure how much a given matrix deviates from a Toeplitz matrix based on the computation of the so-called *displacement rank*. The Toeplitz structure was chosen due to the inherent computational simplicity associated with the algorithms employed to solve Toeplitz-like problems, as described thoroughly in [35, 61].

Let us assume that $\mathbf{A} \in \mathbb{C}^{M_1 \times M_1}$ and $\mathbf{B} \in \mathbb{C}^{M_2 \times M_2}$ are two given square matrices, where M_1 and M_2 are positive integers. From now on, we shall refer to those matrices as *operator matrices*. Consider the linear transformations

$$\begin{aligned}\nabla_{\mathbf{A},\mathbf{B}} : \mathbb{C}^{M_1 \times M_2} &\rightarrow \mathbb{C}^{M_1 \times M_2} \\ \mathbf{C} &\mapsto \nabla_{\mathbf{A},\mathbf{B}}(\mathbf{C}) \triangleq \mathbf{AC} - \mathbf{CB}, \end{aligned} \tag{4.18}$$

$$\begin{aligned}\Delta_{\mathbf{A},\mathbf{B}} : \mathbb{C}^{M_1 \times M_2} &\rightarrow \mathbb{C}^{M_1 \times M_2} \\ \mathbf{C} &\mapsto \Delta_{\mathbf{A},\mathbf{B}}(\mathbf{C}) \triangleq \mathbf{C} - \mathbf{ACB}, \end{aligned} \tag{4.19}$$

which associate each $M_1 \times M_2$ complex-valued rectangular matrix $\mathbf{C}$ with other $M_1 \times M_2$ rectangular matrices $\nabla_{\mathbf{A},\mathbf{B}}(\mathbf{C})$ and $\Delta_{\mathbf{A},\mathbf{B}}(\mathbf{C})$, respectively. These linear mappings are the so-called *Sylvester* ($\nabla_{\mathbf{A},\mathbf{B}}$) and *Stein* ($\Delta_{\mathbf{A},\mathbf{B}}$) *displacement operators*.

The *displacement ranks* related to these linear operators when applied to $\mathbf{C}$ are the ranks of the resulting *displacement matrices* $\nabla_{\mathbf{A},\mathbf{B}}(\mathbf{C})$ and $\Delta_{\mathbf{A},\mathbf{B}}(\mathbf{C})$. In general, matrix $\mathbf{C}$ is represented through $M_1 M_2$ coefficients. However, if $R \triangleq \text{rank}\{\nabla_{\mathbf{A},\mathbf{B}}(\mathbf{C})\}$ or $R \triangleq \text{rank}\{\Delta_{\mathbf{A},\mathbf{B}}(\mathbf{C})\}$ is such that $R \ll \min\{M_1, M_2\}$, then the original matrix can be represented using up to $R(M_1 + M_2) \ll M_1 M_2$ independent coefficients,[8] thus reducing the amount of elements required to express matrix $\mathbf{C}$, as will be further clarified.

The operator matrices $\mathbf{A}$ and $\mathbf{B}$ define completely the displacement operator that will be utilized. The choices of these matrices should rely on which matrices lead to lower displacement ranks. The most common types of operator matrices are the λ-circulant and the diagonal matrices. The *λ-circulant matrix*, denoted as $\mathbf{Z}_\lambda$, and the *diagonal matrix*, denoted as $\mathbf{D}_\nu$, are $M \times M$ square matrices, respectively, given by

$$\mathbf{Z}_\lambda \triangleq \begin{bmatrix} 0 & & & \lambda \\ 1 & \ddots & & \\ & \ddots & \ddots & \\ & & 1 & 0 \end{bmatrix} = \begin{bmatrix} \mathbf{e}_1 & \cdots & \mathbf{e}_{M-1} & \lambda \mathbf{e}_0 \end{bmatrix}, \tag{4.20}$$

$$\mathbf{D}_\nu \triangleq \text{diag}\{\boldsymbol{\nu}\} = \begin{bmatrix} \nu_0 & & & \\ & \nu_1 & & \\ & & \ddots & \\ & & & \nu_{M-1} \end{bmatrix} = \begin{bmatrix} \nu_0 \mathbf{e}_0 & \nu_1 \mathbf{e}_1 & \cdots & \nu_{M-1} \mathbf{e}_{M-1} \end{bmatrix}, \tag{4.21}$$

where λ is a complex-valued scalar, $\mathbf{e}_m$ is an $M \times 1$ vector whose mth entry is equal to 1, whereas the remaining elements are equal to 0, and $\boldsymbol{\nu} \triangleq [\,\nu_0 \ \nu_1 \ \cdots \ \nu_{M-1}\,]^T$ is an $M \times 1$ complex-valued vector. The reader is encouraged to verify that

$$\mathbf{Z}_\lambda^{-1} = \mathbf{Z}_{\frac{1}{\lambda}}^T = \mathbf{Z}_{\frac{1}{\lambda^*}}^H, \tag{4.22}$$

for all non-zero scalar λ and that

$$\mathbf{D}_\nu^{-1} = \text{diag}\left\{\begin{bmatrix} \frac{1}{\nu_0} & \frac{1}{\nu_1} & \cdots & \frac{1}{\nu_{M-1}} \end{bmatrix}^T\right\}, \tag{4.23}$$

as long as ν_m is a non-zero scalar, for each m within the set $\mathcal{M}$ defined as

$$\mathcal{M} \triangleq \{0, 1, \cdots, M-1\}. \tag{4.24}$$

[8] The reader should remember that a rank-1 matrix can be expressed as the outer product between two vectors. In general, a rank-R matrix can be expressed as the sum of R outer products between vectors.

Example 4.2 (λ-Circulant Operator Matrices) This example illustrates the effects of right- and left-multiplying a given matrix by λ-circulant matrices. Consider a 3×3 matrix $\mathbf{C}$ given by

$$\mathbf{C} = \begin{bmatrix} 1 & 2 & 3 \\ 4 & 5 & 6 \\ 7 & 8 & 9 \end{bmatrix}. \tag{4.25}$$

In addition, assume that $\lambda = -1$ and

$$\mathbf{Z}_{-1} = \begin{bmatrix} 0 & 0 & -1 \\ 1 & 0 & 0 \\ 0 & 1 & 0 \end{bmatrix}. \tag{4.26}$$

Hence, we have

$$\mathbf{C}\mathbf{Z}_{-1} = \begin{bmatrix} 2 & 3 & -1 \\ 5 & 6 & -4 \\ 8 & 9 & -7 \end{bmatrix}, \tag{4.27}$$

$$\mathbf{Z}_{-1}\mathbf{C} = \begin{bmatrix} -7 & -8 & -9 \\ 1 & 2 & 3 \\ 4 & 5 & 6 \end{bmatrix}. \tag{4.28}$$

Therefore, right-multiplication by a λ-circulant matrix shifts all columns to the left, where the first original column multiplied by λ is moved to the last column of the resulting matrix. On the other hand, left-multiplying by a λ-circulant matrix shifts down all rows, where the last original row multiplied by λ is moved to the first row of the resulting matrix. □

Usually, structured matrices can be associated with some linear displacement operator. These operators might reveal if a given structured matrix can be represented by a reduced number of parameters. This representation is the key feature that allows the derivation of *superfast algorithms*[9] for inverting as well as performing matrix-to-vector multiplication involving the related structured matrix.

The fast implementation of the reduced-redundancy transceivers rely on the displacement rank of the matrices involved. The procedure entails the following steps.

1. *Compression*: If the rank of the displacement matrix of a given $M_1 \times M_2$ structured matrix $\mathbf{C}$ is lower than the dimensions of $\mathbf{C}$, then it is possible to represent this matrix with a reduced number of coefficients. Indeed, the displacement operator applied to the original matrix can be compressed and represented by the so-called *displacement-generator pair of matrices* $(\mathbf{P}, \mathbf{Q})$ with the following features: assuming we are dealing with a Sylvester displacement operator

[9]That is, algorithms that require $\mathcal{O}(M \log^d M)$ numerical operations, where $d \leq 3$ [61].

of rank $R \in \mathbb{N}$, it is true that

$$\begin{aligned}\nabla_{\mathbf{A},\mathbf{B}}(\mathbf{C}) &= \sum_{r\in\mathcal{R}} \mathbf{p}_r \mathbf{q}_r^H \\ &= \mathbf{P}\mathbf{Q}^H,\end{aligned} \tag{4.29}$$

in which $\mathcal{R} \triangleq \{0, 1, \cdots, R-1\}$ and the generator matrices are given by

$$\mathbf{P} \triangleq \begin{bmatrix}\mathbf{p}_0 & \mathbf{p}_1 & \cdots & \mathbf{p}_{R-1}\end{bmatrix} \in \mathbb{C}^{M_1 \times R}, \tag{4.30}$$

$$\mathbf{Q} \triangleq \begin{bmatrix}\mathbf{q}_0 & \mathbf{q}_1 & \cdots & \mathbf{q}_{R-1}\end{bmatrix} \in \mathbb{C}^{M_2 \times R}, \tag{4.31}$$

where, for each r within the set $\mathcal{R}$, we have

$$\mathbf{p}_r \triangleq \begin{bmatrix}p_{0,r} & p_{1,r} & \cdots & p_{(M_1-1),r}\end{bmatrix}^T, \tag{4.32}$$

$$\mathbf{q}_r \triangleq \begin{bmatrix}q_{0,r} & q_{1,r} & \cdots & q_{(M_2-1),r}\end{bmatrix}^T. \tag{4.33}$$

2. *Operation*: The idea behind the displacement approach is that the compressed form of a structured matrix contains all the information about the original matrix, but with a reduced amount of elements. Therefore, instead of performing operations with the original matrix, it is worth using the appropriate displacement-generator pair to perform such operations.

3. *Decompression*: Once the required operations are performed, the processed version of the original matrix can be recovered through a displacement decompression operation, as long as the operator matrices satisfy some mild constraints.

In order to illustrate the power of the displacement-rank tool, let us define some types of structured matrices in a more formal manner in Subsection 4.3.2.

4.3.2 TOEPLITZ, VANDERMONDE, CAUCHY, AND BEZOUTIAN MATRICES

Toeplitz matrices have the property that their entries do not vary along the diagonals, as can be seen in Equation (4.36). Indeed, an $M_2 \times M_1$ complex-valued matrix $\boldsymbol{T}$ is a *Toeplitz matrix* when, for each pair of indexes (m_2, m_1) within the set $\mathcal{M}_2 \times \mathcal{M}_1$, in which

$$\mathcal{M}_1 \triangleq \{0, 1, \cdots, M_1 - 1\}, \tag{4.34}$$

$$\mathcal{M}_2 \triangleq \{0, 1, \cdots, M_2 - 1\}, \tag{4.35}$$

the (m_2, m_1)th element of $\boldsymbol{T}$, denoted as $[\boldsymbol{T}]_{m_2 m_1} \triangleq t_{m_2,m_1} \in \mathbb{C}$, satisfies the equality $t_{m_2,m_1} = t_{(m_2+1),(m_1+1)}$, as long as the pair of indexes (m_2+1, m_1+1) is also within the set $\mathcal{M}_2 \times \mathcal{M}_1$. In this case, matrix $\boldsymbol{T}$ can be rewritten in a more convenient form, as follows:

$$\boldsymbol{T} = \begin{bmatrix} t_0 & t_{-1} & t_{-2} & \cdots & t_{1-M_1} \\ t_1 & t_0 & t_{-1} & \cdots & t_{2-M_1} \\ t_2 & t_1 & t_0 & \cdots & t_{3-M_1} \\ \vdots & \vdots & \vdots & \vdots & \vdots \\ t_{M_2-1} & t_{M_2-2} & t_{M_2-3} & \cdots & t_{M_2-M_1} \end{bmatrix} \triangleq \left[t_{(m_2-m_1)}\right]_{(m_2,m_1)\in\mathcal{M}_2\times\mathcal{M}_1}. \tag{4.36}$$

In order to exemplify the compression capability of the displacement operators when dealing with rectangular Toeplitz matrices, let us consider the application of the Sylvester displacement operator $\nabla_{\mathbf{Z}_\eta,\mathbf{Z}_\xi}$, in which $\mathbf{Z}_\eta \in \mathbb{C}^{M_2\times M_2}$ and $\mathbf{Z}_\xi \in \mathbb{C}^{M_1\times M_1}$, on an $M_2 \times M_1$ complex-valued Toeplitz matrix $\boldsymbol{T}$:

$$\begin{aligned} \nabla_{\mathbf{Z}_\eta,\mathbf{Z}_\xi}(\boldsymbol{T}) &= \mathbf{Z}_\eta \boldsymbol{T} - \boldsymbol{T}\mathbf{Z}_\xi \\ &= \begin{bmatrix} \eta t_{M_2-1} & \eta t_{M_2-2} & \cdots & \eta t_{M_2-M_1+1} & \eta t_{M_2-M_1} \\ t_0 & t_{-1} & \cdots & t_{2-M_1} & t_{1-M_1} \\ t_1 & t_0 & \cdots & t_{3-M_1} & t_{2-M_1} \\ \vdots & \vdots & \vdots & \vdots & \vdots \\ t_{M_2-2} & t_{M_2-3} & \cdots & t_{M_2-M_1} & t_{M_2-M_1-1} \end{bmatrix} \\ &\quad - \begin{bmatrix} t_{-1} & t_{-2} & \cdots & t_{1-M_1} & \xi t_0 \\ t_0 & t_{-1} & \cdots & t_{2-M_1} & \xi t_1 \\ \vdots & \vdots & \vdots & \vdots & \vdots \\ t_{M_2-3} & t_{M_2-4} & \cdots & t_{M_2-M_1-1} & \xi t_{M_2-2} \\ t_{M_2-2} & t_{M_2-3} & \cdots & t_{M_2-M_1} & \xi t_{M_2-1} \end{bmatrix} \\ &= \begin{bmatrix} \eta t_{M_2-1} - t_{-1} & \eta t_{M_2-2} - t_{-2} & \cdots & \eta t_{M_2-M_1+1} - t_{1-M_1} & \eta t_{M_2-M_1} - \xi t_0 \\ 0 & 0 & \cdots & 0 & t_{1-M_1} - \xi t_1 \\ \vdots & \vdots & \vdots & \vdots & \vdots \\ 0 & 0 & \cdots & 0 & t_{M_2-M_1-2} - \xi t_{M_2-2} \\ 0 & 0 & \cdots & 0 & t_{M_2-M_1-1} - \xi t_{M_2-1} \end{bmatrix} \end{aligned}$$

$$
\begin{aligned}
&= \underbrace{\begin{bmatrix} 1 \\ 0 \\ \vdots \\ 0 \\ 0 \end{bmatrix}}_{\triangleq \hat{\mathbf{p}}_1} \underbrace{\begin{bmatrix} \eta t_{M_2-1} - t_{-1} & \eta t_{M_2-2} - t_{-2} & \cdots & \eta t_{M_2-M_1+1} - t_{1-M_1} & \eta t_{M_2-M_1} \end{bmatrix}}_{\triangleq \hat{\mathbf{q}}_1^H} \\
&\quad + \underbrace{\begin{bmatrix} -\xi t_0 \\ t_{1-M_1} - \xi t_1 \\ \vdots \\ t_{M_2-M_1-2} - \xi t_{M_2-2} \\ t_{M_2-M_1-1} - \xi t_{M_2-1} \end{bmatrix}}_{\triangleq \hat{\mathbf{p}}_2} \underbrace{\begin{bmatrix} 0 & 0 & \cdots & 0 & 1 \end{bmatrix}}_{\triangleq \hat{\mathbf{q}}_2^H} \\
&= \hat{\mathbf{p}}_1 \hat{\mathbf{q}}_1^H + \hat{\mathbf{p}}_2 \hat{\mathbf{q}}_2^H \\
&= \underbrace{\begin{bmatrix} \hat{\mathbf{p}}_1 & \hat{\mathbf{p}}_2 \end{bmatrix}}_{\triangleq \hat{\mathbf{P}}} \underbrace{\begin{bmatrix} \hat{\mathbf{q}}_1^H \\ \hat{\mathbf{q}}_2^H \end{bmatrix}}_{\triangleq \hat{\mathbf{Q}}^H} \\
&= \hat{\mathbf{P}} \hat{\mathbf{Q}}^H .
\end{aligned}
\tag{4.37}
$$

Note that the resulting displacement matrix $\nabla_{\mathbf{Z}_\eta, \mathbf{Z}_\xi}(\boldsymbol{T})$ can be represented by the displacement-generator pair of matrices $(\hat{\mathbf{P}}, \hat{\mathbf{Q}}) \in \mathbb{C}^{M_2 \times 2} \times \mathbb{C}^{M_1 \times 2}$. Thus, if one assumes that M_2 and M_1 are integer numbers much larger than 2, then the former example shows that rectangular Toeplitz matrices can always be compressed, since matrix $\nabla_{\mathbf{Z}_\eta, \mathbf{Z}_\xi}(\boldsymbol{T})$ has rank at most 2. In fact, since the vectors $\hat{\mathbf{p}}_1$ and $\hat{\mathbf{q}}_2$ do not depend on $\boldsymbol{T}$, then at most $M_2 + M_1$ coefficients (corresponding to the elements of the vectors $\hat{\mathbf{p}}_2$ and $\hat{\mathbf{q}}_1$) are really required to represent the displacement of $\boldsymbol{T}$. This reduced amount of coefficients is rather close to the total number of possibly different entries of matrix $\boldsymbol{T}$, which is $M_2 + M_1 - 1$.

Now, let us define another important class of structured matrix: the Vandermonde matrix, which is a matrix whose entries in each row are the terms of a geometric progression, as can be seen in Equation (4.38) below. Indeed, given an $M \times 1$ complex-valued vector $\boldsymbol{\nu} \triangleq [\, \nu_0 \; \nu_1 \; \cdots \; \nu_{M-1} \,]^T$, an $M \times M$ matrix $\boldsymbol{V}_{\boldsymbol{\nu}}$ is a *Vandermonde matrix* when, for each pair of indexes (m_2, m_1) within the set $\mathcal{M}^2 \triangleq \mathcal{M} \times \mathcal{M}$, the (m_2, m_1)th element of $\boldsymbol{V}_{\boldsymbol{\nu}}$, denoted as $[\boldsymbol{V}_{\boldsymbol{\nu}}]_{m_2 m_1} \triangleq v_{m_2, m_1} \in \mathbb{C}$, satisfies the equality $\nu_{m_2,m_1} = \nu_{m_2}^{m_1}$. In this case, $\boldsymbol{V}_{\boldsymbol{\nu}}$ presents the following form:

$$\begin{aligned}\boldsymbol{V}_{\boldsymbol{\nu}} &= \begin{bmatrix} \nu_0^0 & \nu_0^1 & \cdots & \nu_0^{M-2} & \nu_0^{M-1} \\ \nu_1^0 & \nu_1^1 & \cdots & \nu_1^{M-2} & \nu_1^{M-1} \\ \vdots & \vdots & \vdots & \vdots & \vdots \\ \nu_{M-2}^0 & \nu_{M-2}^1 & \cdots & \nu_{M-2}^{M-2} & \nu_{M-2}^{M-1} \\ \nu_{M-1}^0 & \nu_{M-1}^1 & \cdots & \nu_{M-1}^{M-2} & \nu_{M-1}^{M-1} \end{bmatrix} \\ &\triangleq [\nu_{m_2}^{m_1}]_{(m_2,m_1)\in\mathcal{M}^2}. \end{aligned} \tag{4.38}$$

As in the Toeplitz case (see Equation (4.37)), a similar kind of compression can be applied to Vandermonde matrices as well. Let us consider the application of the Sylvester displacement operator $\nabla_{\mathbf{D}_{\boldsymbol{\nu}},\mathbf{Z}_0}$, in which $\mathbf{D}_{\boldsymbol{\nu}} = \operatorname{diag}\{\boldsymbol{\nu}\} \in \mathbb{C}^{M\times M}$, on a given $M \times M$ Vandermonde matrix $\boldsymbol{V}_{\boldsymbol{\nu}}$. In this case, we have

$$\begin{aligned}\nabla_{\mathbf{D}_{\boldsymbol{\nu}},\mathbf{Z}_0}(\boldsymbol{V}_{\boldsymbol{\nu}}) &= \mathbf{D}_{\boldsymbol{\nu}}\boldsymbol{V}_{\boldsymbol{\nu}} - \boldsymbol{V}_{\boldsymbol{\nu}}\mathbf{Z}_0 \\ &= \begin{bmatrix} \nu_0 & \cdots & \nu_0^{M-1} & \nu_0^M \\ \nu_1 & \cdots & \nu_1^{M-1} & \nu_1^M \\ \vdots & & \vdots & \vdots \\ \nu_{M-1} & \cdots & \nu_{M-1}^{M-1} & \nu_{M-1}^M \end{bmatrix} - \begin{bmatrix} \nu_0 & \cdots & \nu_0^{M-1} & 0 \\ \nu_1 & \cdots & \nu_1^{M-1} & 0 \\ \vdots & & \vdots & \vdots \\ \nu_{M-1} & \cdots & \nu_{M-1}^{M-1} & 0 \end{bmatrix} \\ &= \begin{bmatrix} 0 & \cdots & 0 & \nu_0^M \\ 0 & \cdots & 0 & \nu_1^M \\ \vdots & & \vdots & \vdots \\ 0 & \cdots & 0 & \nu_{M-1}^M \end{bmatrix}, \end{aligned} \tag{4.39}$$

which consists of a rank-1 matrix with M degrees of freedom. Note that, even though the original Vandermonde matrix $\boldsymbol{V}_{\boldsymbol{\nu}}$ is comprised of M^2 entries, the M elements which compose the vector $\boldsymbol{\nu}$ are enough to completely define $\boldsymbol{V}_{\boldsymbol{\nu}}$. This compression example, therefore, shows that the displacement approach is able to reveal analytically this reduced number of degrees of freedom.

In addition, there is a close relation between Vandermonde and DFT matrices. In order to derive such a useful relation, let us first remember that the Mth roots of a given complex number ξ consist of M distinct complex numbers ξ_m, with $m \in \mathcal{M}$, such that $\xi_m^M = \xi = |\xi|e^{j\angle\xi}$, where $j^2 = -1$ and $\angle\xi \in (-\pi, \pi] \subset \mathbb{R}$ represents the (principal)[10] phase of the complex number ξ when

[10]Remember that, if $\angle\xi$ is a phase of a given complex number ξ, then $\angle\xi + 2i\pi$ is also a phase of ξ, for any integer number i. The *principal phase* is the unique phase of ξ within the interval $(-\pi, \pi]$.

expressed in its polar form; that is,

$$\begin{aligned}\xi_m &\triangleq \underbrace{|\xi|^{\frac{1}{M}} \mathrm{e}^{\mathrm{j}\frac{\angle \xi}{M}}}_{\triangleq \xi_0} \mathrm{e}^{-\mathrm{j}\frac{2\pi}{M}m} \\ &= \xi_0 \Bigg(\underbrace{\mathrm{e}^{-\mathrm{j}\frac{2\pi}{M}}}_{\triangleq W_M}\Bigg)^m \\ &= \xi_0 W_M^m. \end{aligned} \tag{4.40}$$

Therefore, by defining the vector $\boldsymbol{\xi}$ as

$$\boldsymbol{\xi} \triangleq \begin{bmatrix}\xi_0 & \xi_1 & \cdots & \xi_{M-1}\end{bmatrix}^T, \tag{4.41}$$

its corresponding Vandermonde matrix is given by

$$\begin{aligned}\boldsymbol{V}_\xi &= \begin{bmatrix} \xi_0^0 & \xi_0^1 & \cdots & \xi_0^{M-1} \\ \xi_1^0 & \xi_1^1 & \cdots & \xi_1^{M-1} \\ \vdots & \vdots & \vdots & \vdots \\ \xi_{M-1}^0 & \xi_{M-1}^1 & \cdots & \xi_{M-1}^{M-1} \end{bmatrix} \\ &= \begin{bmatrix} \xi_0^0 W_M^0 & \xi_0^1 W_M^0 & \cdots & \xi_0^{M-1} W_M^0 \\ \xi_0^0 W_M^0 & \xi_0^1 W_M^1 & \cdots & \xi_0^{M-1} W_M^{M-1} \\ \vdots & \vdots & \vdots & \vdots \\ \xi_0^0 W_M^0 & \xi_0^1 W_M^{M-1} & \cdots & \xi_0^{M-1} W_M^{(M-1)^2} \end{bmatrix} \\ &= \begin{bmatrix} W_M^0 & W_M^0 & \cdots & W_M^0 \\ W_M^0 & W_M^1 & \cdots & W_M^{M-1} \\ \vdots & \vdots & \vdots & \vdots \\ W_M^0 & W_M^{M-1} & \cdots & W_M^{(M-1)^2} \end{bmatrix} \begin{bmatrix} \xi_0^0 & 0 & \cdots & 0 \\ 0 & \xi_0^1 & \cdots & 0 \\ \vdots & \vdots & \vdots & \vdots \\ 0 & 0 & \cdots & \xi_0^{M-1} \end{bmatrix} \\ &= \sqrt{M}\mathbf{W}_M \operatorname{diag}\{\xi_0^m\}_{m\in\mathcal{M}}, \end{aligned} \tag{4.42}$$

where $\mathbf{W}_M$ is the $M \times M$ unitary DFT matrix previously defined in Equation (2.46). Such a relation between Vandermonde and DFT matrices will be key to developing efficient factorizations of structured matrices which appear in the ZP-ZJ model.

Another important structured matrix is the so-called Cauchy matrix, which may not be familiar to some readers. Given two vectors, $\boldsymbol{\nu} \triangleq \begin{bmatrix}\nu_0 & \nu_1 & \cdots & \nu_{M_1-1}\end{bmatrix}^T \in \mathbb{C}^{M_1\times 1}$ and $\boldsymbol{\lambda} \triangleq \begin{bmatrix}\lambda_0 & \lambda_1 & \cdots & \lambda_{M_2-1}\end{bmatrix}^T \in \mathbb{C}^{M_2\times 1}$, an $M_1 \times M_2$ complex-valued matrix $\boldsymbol{C}_{\nu,\lambda}$ is a *Cauchy matrix* when, for any pair of indexes (m_1, m_2) within the set $\mathcal{M}_1 \times \mathcal{M}_2$, the (m_1, m_2)th element of $\boldsymbol{C}_{\nu,\lambda}$, denoted as $\left[\boldsymbol{C}_{\nu,\lambda}\right]_{m_1 m_2} \triangleq c_{m_1,m_2} \in \mathbb{C}$, satisfies the equality $\frac{1}{c_{m_1,m_2}} = 1 - \nu_{m_1}\lambda_{m_2}$, assuming that

$\nu_{m_1}\lambda_{m_2} \neq 1$. Thus, $\boldsymbol{C}_{\nu,\lambda}$ can be represented as follows:

$$\boldsymbol{C}_{\nu,\lambda} = \begin{bmatrix} \frac{1}{1-\nu_0\lambda_0} & \frac{1}{1-\nu_0\lambda_1} & \cdots & \frac{1}{1-\nu_0\lambda_{(M_2-1)}} \\ \frac{1}{1-\nu_1\lambda_0} & \frac{1}{1-\nu_1\lambda_1} & \cdots & \frac{1}{1-\nu_1\lambda_{(M_2-1)}} \\ \vdots & \vdots & \ddots & \vdots \\ \frac{1}{1-\nu_{(M_1-1)}\lambda_0} & \frac{1}{1-\nu_{(M_1-1)}\lambda_1} & \cdots & \frac{1}{1-\nu_{(M_1-1)}\lambda_{(M_2-1)}} \end{bmatrix}$$

$$\triangleq \left[\frac{1}{1-\nu_{m_1}\lambda_{m_2}}\right]_{(m_1,m_2)\in\mathcal{M}_1\times\mathcal{M}_2}. \tag{4.43}$$

A Cauchy matrix $\boldsymbol{C}_{\nu,\lambda}$ can be easily compressed by using the Stein displacement operator $\Delta_{\mathbf{D}_\nu,\mathbf{D}_\lambda}$. Indeed, one has

$$\begin{aligned} \Delta_{\mathbf{D}_\nu,\mathbf{D}_\lambda}(\boldsymbol{C}_{\nu,\lambda}) &= \boldsymbol{C}_{\nu,\lambda} - \mathbf{D}_\nu \boldsymbol{C}_{\nu,\lambda}\mathbf{D}_\lambda \\ &= \left[\frac{1}{1-\nu_{m_1}\lambda_{m_2}}\right]_{(m_1,m_2)\in\mathcal{M}_1\times\mathcal{M}_2} - \left[\frac{\nu_{m_1}\lambda_{m_2}}{1-\nu_{m_1}\lambda_{m_2}}\right]_{(m_1,m_2)\in\mathcal{M}_1\times\mathcal{M}_2} \\ &= \left[\frac{1-\nu_{m_1}\lambda_{m_2}}{1-\nu_{m_1}\lambda_{m_2}}\right]_{(m_1,m_2)\in\mathcal{M}_1\times\mathcal{M}_2} \\ &= \begin{bmatrix} 1 & \cdots & 1 \\ \vdots & \ddots & \vdots \\ 1 & \cdots & 1 \end{bmatrix}, \end{aligned} \tag{4.44}$$

which is also a rank-1 matrix. In fact, the resulting displacement matrix in Equation (4.44) enjoys the most simple form that we have found up to now. This feature will be important for us in Subsection 4.4.1. Moreover, a rectangular matrix $\boldsymbol{C}^{g}_{\nu,\lambda}$ is defined as a *generalized Cauchy matrix*, as long as the Stein displacement matrix $\Delta_{\mathbf{D}_\nu,\mathbf{D}_\lambda}(\boldsymbol{C}^{g}_{\nu,\lambda})$ has rank much smaller than the number of rows and columns of $\boldsymbol{C}^{g}_{\nu,\lambda}$.[11]

The last class of structured matrix of our interest is the Bezoutian matrix. As opposed to most of the previous structured matrices, there is no simple closed-form expression for the entries of a Bezoutian matrix. In fact, the easiest way to define a Bezoutian matrix is based on its displacement rank, as in the case of generalized Cauchy matrices. Indeed, an $M_1 \times M_2$ *Bezoutian matrix* $\boldsymbol{B}$ is such that

$$\nabla_{\mathbf{Z}_\nu,\mathbf{Z}_\lambda}(\boldsymbol{B}) = \mathbf{P}\mathbf{Q}^H \tag{4.45}$$

for some complex-valued scalars ν and λ, where $(\mathbf{P},\mathbf{Q}) \in \mathbb{C}^{M_1\times R} \times \mathbb{C}^{M_2\times R}$, and with $M_2 \gg R$ and $M_1 \gg R$.

It is possible to show (see Subsection 4.3.3) that the inverse of a non-singular square Toeplitz matrix is a particular type of Bezoutian matrix, also known as *T-Bezoutian matrix*.

[11]There are some other classes of Cauchy matrices which will not be of particular importance in this book. The interested reader should refer to [27, 28] for detailed information on this topic.

So far, we have only illustrated the compression capabilities of the displacement operators. Let us analyze now the operation stage of the displacement approach.

4.3.3 PROPERTIES OF DISPLACEMENT OPERATORS

As we already mentioned, the compression stage is key to representing a given structured matrix using a reduced amount of elements. Nevertheless, in order to benefit from the reduced amount of elements in practice, it is necessary to conceive some ways of translating the operations on the compressed form into related operations over the original matrix entries. The operation stage associated with the displacement approach is in charge of such a translation. Thus, this subsection addresses some properties which describe how standard matrix operations, namely linear combination, product, and inversion, are mapped into their corresponding displacement matrices.

A first key result related to the displacement operators is the conversion between the Sylvester and Stein operators. Such a conversion is possible as long as at least one of the operator matrices, either $\mathbf{A}$ or $\mathbf{B}$ (see expressions (4.18) and (4.19)), is invertible. For example, if the operator matrix $\mathbf{A}$ is a square non-singular matrix, then

$$\begin{aligned}\nabla_{\mathbf{A},\mathbf{B}}(\mathbf{C}) &= \mathbf{AC} - \mathbf{CB} \\ &= \mathbf{A}\left(\mathbf{C} - \mathbf{A}^{-1}\mathbf{CB}\right) \\ &= \mathbf{A}\Delta_{\mathbf{A}^{-1},\mathbf{B}}(\mathbf{C}),\end{aligned} \tag{4.46}$$

in which it is assumed that all matrices have compatible dimensions.[12] Similarly, if $\mathbf{B}$ is invertible, then

$$\nabla_{\mathbf{A},\mathbf{B}}(\mathbf{C}) = -\Delta_{\mathbf{A},\mathbf{B}^{-1}}(\mathbf{C})\mathbf{B}. \tag{4.47}$$

These relations between the Sylvester and Stein displacement operators show that, given two operator matrices $\mathbf{A}$ and $\mathbf{B}$, the displacement rank of a third matrix $\mathbf{C}$ does not depend on the particular choice of the displacement operator (whether Sylvester or Stein), as long as at least one of the operator matrices has full-rank and, in addition, some minor adjustments on the operator matrices are performed, as for example, replacing $\mathbf{A}$ by $\mathbf{A}^{-1}$ or $\mathbf{B}$ by $\mathbf{B}^{-1}$.

Let us now discuss two important results dealing with linear combinations of matrices and matrix products, which basically show how these operations affect the displacement-generator pairs of the original matrices. With respect to the generator of a linear combination, assume that $\nabla_{\mathbf{A},\mathbf{B}}(\mathbf{C}) = \mathbf{P}_{\mathbf{C}}\mathbf{Q}_{\mathbf{C}}^{H}$ and that $\nabla_{\mathbf{A},\mathbf{B}}(\mathbf{D}) = \mathbf{P}_{\mathbf{D}}\mathbf{Q}_{\mathbf{D}}^{H}$, where matrices $\mathbf{C}$ and $\mathbf{D}$ have the same dimen-

[12] That is, the number of rows and columns are such that one can perform operations with the involved matrices, like addition or product, without worrying about the dimensions of the related matrices.

sions. Thus, for any complex-valued scalar α, one always has

$$\begin{aligned}
\nabla_{\mathbf{A},\mathbf{B}}(\mathbf{C}+\alpha\mathbf{D}) &= \nabla_{\mathbf{A},\mathbf{B}}(\mathbf{C})+\alpha\nabla_{\mathbf{A},\mathbf{B}}(\mathbf{D}) \\
&= \mathbf{P}_{\mathbf{C}}\mathbf{Q}_{\mathbf{C}}^{H}+\alpha\mathbf{P}_{\mathbf{D}}\mathbf{Q}_{\mathbf{D}}^{H} \\
&= \underbrace{\begin{bmatrix}\mathbf{P}_{\mathbf{C}} & \alpha\mathbf{P}_{\mathbf{D}}\end{bmatrix}}_{\triangleq\mathbf{P}}\underbrace{\begin{bmatrix}\mathbf{Q}_{\mathbf{C}}^{H} \\ \mathbf{Q}_{\mathbf{D}}^{H}\end{bmatrix}}_{\triangleq\mathbf{Q}^{H}} \\
&= \mathbf{P}\mathbf{Q}^{H}.
\end{aligned} \tag{4.48}$$

As for the generator of a product, consider that $\nabla_{\mathbf{A},\mathbf{B}}(\mathbf{C}) = \mathbf{P}_{\mathbf{C}}\mathbf{Q}_{\mathbf{C}}^{H}$ and that $\nabla_{\mathbf{B},\mathbf{D}}(\mathbf{E}) = \mathbf{P}_{\mathbf{E}}\mathbf{Q}_{\mathbf{E}}^{H}$, where all matrices have compatible dimensions. Hence, we have

$$\begin{aligned}
\nabla_{\mathbf{A},\mathbf{D}}(\mathbf{CE}) &= \mathbf{A}(\mathbf{CE})-(\mathbf{CE})\mathbf{D} \\
&= (\mathbf{ACE}-\mathbf{CBE})+(\mathbf{CBE}-\mathbf{CED}) \\
&= (\mathbf{AC}-\mathbf{CB})\,\mathbf{E}+\mathbf{C}\,(\mathbf{BE}-\mathbf{ED}) \\
&= \left(\mathbf{P}_{\mathbf{C}}\mathbf{Q}_{\mathbf{C}}^{H}\right)\mathbf{E}+\mathbf{C}\left(\mathbf{P}_{\mathbf{E}}\mathbf{Q}_{\mathbf{E}}^{H}\right) \\
&= \underbrace{\begin{bmatrix}\mathbf{P}_{\mathbf{C}} & \mathbf{C}\mathbf{P}_{\mathbf{E}}\end{bmatrix}}_{\triangleq\mathbf{P}}\underbrace{\begin{bmatrix}\mathbf{Q}_{\mathbf{C}}^{H}\mathbf{E} \\ \mathbf{Q}_{\mathbf{E}}^{H}\end{bmatrix}}_{\triangleq\mathbf{Q}^{H}} \\
&= \mathbf{P}\mathbf{Q}^{H}.
\end{aligned} \tag{4.49}$$

Thus, the knowledge about the displacement-generator pairs of the individual matrices before the algebraic operation (whether linear combination or product) helps one determine the displacement-generator pair of the resulting matrix after the related operation.

Now, we will analyze the inversion operation. The importance of this result is the fact that the compression of an inverse matrix can be easily generated from the compressed version of the original matrix. Indeed, given an invertible square complex-valued matrix $\mathbf{C}$ and the operator matrices $\mathbf{A}$ and $\mathbf{B}$, one has

$$\begin{aligned}
\nabla_{\mathbf{B},\mathbf{A}}(\mathbf{C}^{-1}) &= \mathbf{B}\mathbf{C}^{-1}-\mathbf{C}^{-1}\mathbf{A} \\
&= \left(\mathbf{B}-\mathbf{C}^{-1}\mathbf{A}\mathbf{C}\right)\mathbf{C}^{-1} \\
&= \mathbf{C}^{-1}\left(\mathbf{CB}-\mathbf{AC}\right)\mathbf{C}^{-1} \\
&= -\mathbf{C}^{-1}\nabla_{\mathbf{A},\mathbf{B}}(\mathbf{C})\mathbf{C}^{-1}.
\end{aligned} \tag{4.50}$$

In other words, if the displacement-generator pair of $\nabla_{\mathbf{A},\mathbf{B}}(\mathbf{C})$ is $(\mathbf{P}_{\mathbf{C}}, \mathbf{Q}_{\mathbf{C}})$, then the displacement-generator pair of $\nabla_{\mathbf{B},\mathbf{A}}(\mathbf{C}^{-1})$ is $(-\mathbf{C}^{-1}\mathbf{P}_{\mathbf{C}}, \mathbf{C}^{-H}\mathbf{Q}_{\mathbf{C}})$, in which $\mathbf{C}^{-H}$ denotes the Hermitian of the inverse of $\mathbf{C}$.

An immediate application of the inversion property is the verification of the following fact: for a given non-singular square Toeplitz matrix $\boldsymbol{T} \in \mathbb{C}^{M\times M}$ whose displacement generator pair associated with the Sylvester displacement operator $\nabla_{\mathbf{Z}_\eta,\mathbf{Z}_\xi}$ is $(\hat{\mathbf{P}},\hat{\mathbf{Q}})$, one has $\nabla_{\mathbf{Z}_\xi,\mathbf{Z}_\eta}(\boldsymbol{T}^{-1}) = (-\boldsymbol{T}^{-1}\hat{\mathbf{P}})(\hat{\mathbf{Q}}^H\boldsymbol{T}^{-1})$, whose rank is at most 2 due to Equation (4.37). Therefore, $\boldsymbol{T}^{-1}$ is a Bezoutian matrix whenever $M \gg 2$.

Actually, any regularized pseudo-inverse of an $M_2 \times M_1$ complex-valued Toeplitz matrix $\boldsymbol{T}$ is a Bezoutian matrix, as long as $\min\{M_2, M_1\} \gg 4$. Indeed, given a strictly positive real number ρ, the *regularized pseudo-inverse* of $\boldsymbol{T}$ is defined as

$$\boldsymbol{T}_\rho^\dagger \triangleq \left(\boldsymbol{T}^H\boldsymbol{T} + \rho\mathbf{I}_{M_1}\right)^{-1}\boldsymbol{T}^H. \tag{4.51}$$

Note that, by applying the *matrix inversion lemma*[13] to the expression $\left(\boldsymbol{T}^H\boldsymbol{T} + \rho\mathbf{I}_{M_1}\right)^{-1}$ that appears in Equation (4.51), one has

$$\begin{aligned}\left(\boldsymbol{T}^H\boldsymbol{T} + \rho\mathbf{I}_{M_1}\right)^{-1} &= \frac{1}{\rho}\mathbf{I}_{M_1} - \frac{1}{\rho}\boldsymbol{T}^H\left(\frac{1}{\rho}\boldsymbol{T}\boldsymbol{T}^H + \mathbf{I}_{M_2}\right)^{-1}\boldsymbol{T}\frac{1}{\rho}\\ &= \frac{1}{\rho}\left[\mathbf{I}_{M_1} - \boldsymbol{T}^H\left(\boldsymbol{T}\boldsymbol{T}^H + \rho\mathbf{I}_{M_2}\right)^{-1}\boldsymbol{T}\right],\end{aligned} \tag{4.52}$$

and, by right-multiplying both sides of Equation (4.52) by matrix $\boldsymbol{T}^H$, we get

$$\begin{aligned}\boldsymbol{T}_\rho^\dagger &= \frac{1}{\rho}\left[\boldsymbol{T}^H - \boldsymbol{T}^H\left(\boldsymbol{T}\boldsymbol{T}^H + \rho\mathbf{I}_{M_2}\right)^{-1}\boldsymbol{T}\boldsymbol{T}^H\right]\\ &= \frac{1}{\rho}\boldsymbol{T}^H\left[\mathbf{I}_{M_2} - \left(\boldsymbol{T}\boldsymbol{T}^H + \rho\mathbf{I}_{M_2}\right)^{-1}\boldsymbol{T}\boldsymbol{T}^H\right]\\ &= \frac{1}{\rho}\boldsymbol{T}^H\left(\boldsymbol{T}\boldsymbol{T}^H + \rho\mathbf{I}_{M_2}\right)^{-1}\left[\left(\boldsymbol{T}\boldsymbol{T}^H + \rho\mathbf{I}_{M_2}\right) - \boldsymbol{T}\boldsymbol{T}^H\right]\\ &= \frac{1}{\rho}\boldsymbol{T}^H\left(\boldsymbol{T}\boldsymbol{T}^H + \rho\mathbf{I}_{M_2}\right)^{-1}\left(\rho\mathbf{I}_{M_2}\right)\\ &= \boldsymbol{T}^H\left(\boldsymbol{T}\boldsymbol{T}^H + \rho\mathbf{I}_{M_2}\right)^{-1}.\end{aligned} \tag{4.53}$$

[13]The following identity holds: $\left(\mathbf{A} + \mathbf{B}\mathbf{C}\mathbf{D}^H\right)^{-1} = \mathbf{A}^{-1} - \mathbf{A}^{-1}\mathbf{B}\left(\mathbf{C}^{-1} + \mathbf{D}^H\mathbf{A}^{-1}\mathbf{B}\right)^{-1}\mathbf{D}^H\mathbf{A}^{-1}$, assuming that all operations are valid (see, for example, [83]).

Therefore, the Sylvester displacement of the regularized pseudo-inverse described in Equations (4.51) and (4.53) is given by

$$\begin{aligned}
\nabla_{\mathbf{Z}_\xi,\mathbf{Z}_\eta}\left(\boldsymbol{T}_\rho^\dagger\right) &= \mathbf{Z}_\xi \boldsymbol{T}^H\left(\boldsymbol{T}\boldsymbol{T}^H+\rho\mathbf{I}_{M_2}\right)^{-1} - \left(\boldsymbol{T}^H\boldsymbol{T}+\rho\mathbf{I}_{M_1}\right)^{-1}\boldsymbol{T}^H\mathbf{Z}_\eta \\
&= \left(\boldsymbol{T}^H\boldsymbol{T}+\rho\mathbf{I}_{M_1}\right)^{-1}\left[\left(\boldsymbol{T}^H\boldsymbol{T}+\rho\mathbf{I}_{M_1}\right)\mathbf{Z}_\xi\boldsymbol{T}^H\right. \\
&\quad \left. - \boldsymbol{T}^H\mathbf{Z}_\eta\left(\boldsymbol{T}\boldsymbol{T}^H+\rho\mathbf{I}_{M_2}\right)\right]\left(\boldsymbol{T}\boldsymbol{T}^H+\rho\mathbf{I}_{M_2}\right)^{-1} \\
&= \left(\boldsymbol{T}^H\boldsymbol{T}+\rho\mathbf{I}_{M_1}\right)^{-1}\left[\left(\boldsymbol{T}^H\boldsymbol{T}\mathbf{Z}_\xi\boldsymbol{T}^H - \boldsymbol{T}^H\mathbf{Z}_\eta\boldsymbol{T}\boldsymbol{T}^H\right)\right. \\
&\quad \left. + \rho\left(\mathbf{Z}_\xi\boldsymbol{T}^H - \boldsymbol{T}^H\mathbf{Z}_\eta\right)\right]\left(\boldsymbol{T}\boldsymbol{T}^H+\rho\mathbf{I}_{M_2}\right)^{-1} \\
&= \left(\boldsymbol{T}^H\boldsymbol{T}+\rho\mathbf{I}_{M_1}\right)^{-1}\left[-\boldsymbol{T}^H\nabla_{\mathbf{Z}_\eta,\mathbf{Z}_\xi}(\boldsymbol{T})\boldsymbol{T}^H\right. \\
&\quad \left. + \rho\nabla_{\mathbf{Z}_\xi,\mathbf{Z}_\eta}\left(\boldsymbol{T}^H\right)\right]\left(\boldsymbol{T}\boldsymbol{T}^H+\rho\mathbf{I}_{M_2}\right)^{-1}.
\end{aligned} \tag{4.54}$$

Hence, if we define $\nabla_{\mathbf{Z}_\eta,\mathbf{Z}_\xi}(\boldsymbol{T})$ as $\hat{\mathbf{P}}\hat{\mathbf{Q}}^H$ and $\nabla_{\mathbf{Z}_\xi,\mathbf{Z}_\eta}\left(\boldsymbol{T}^H\right)$ as $\check{\mathbf{P}}\check{\mathbf{Q}}^H$, then we have

$$\begin{aligned}
\nabla_{\mathbf{Z}_\xi,\mathbf{Z}_\eta}\left(\boldsymbol{T}_\rho^\dagger\right) &= -\boldsymbol{T}_\rho^\dagger\hat{\mathbf{P}}\hat{\mathbf{Q}}^H\boldsymbol{T}_\rho^\dagger + \left(\boldsymbol{T}^H\boldsymbol{T}+\rho\mathbf{I}_{M_1}\right)^{-1}\check{\mathbf{P}}\check{\mathbf{Q}}^H\rho\left(\boldsymbol{T}\boldsymbol{T}^H+\rho\mathbf{I}_{M_2}\right)^{-1} \\
&= \underbrace{\left[\left(\boldsymbol{T}^H\boldsymbol{T}+\rho\mathbf{I}_{M_1}\right)^{-1}\check{\mathbf{P}} \quad -\boldsymbol{T}_\rho^\dagger\hat{\mathbf{P}}\right]}_{\triangleq \mathbf{P}\in\mathbb{C}^{M_1\times 4}}\underbrace{\begin{bmatrix}\check{\mathbf{Q}}^H\rho\left(\boldsymbol{T}\boldsymbol{T}^H+\rho\mathbf{I}_{M_2}\right)^{-1} \\ \hat{\mathbf{Q}}^H\boldsymbol{T}_\rho^\dagger\end{bmatrix}}_{\triangleq \mathbf{Q}^H\in\mathbb{C}^{4\times M_2}} \\
&= \mathbf{P}\mathbf{Q}^H,
\end{aligned} \tag{4.55}$$

which proves that $\boldsymbol{T}_\rho^\dagger$ is a Bezoutian matrix if one assumes that $\min\{M_2, M_1\} \gg 4$.

The above derivation considers that $\rho > 0$. In fact, if we assume that $M_2 \geq M_1$ and that $\operatorname{rank}\{\boldsymbol{T}\} = M_1$, then the $M_2 \times M_2$ matrix $\boldsymbol{T}\boldsymbol{T}^H$ turns out to be rank deficient. In such a case, the previous derivation would not be valid for $\rho \to 0$, due to the use of operations like the inversion of $\left(\boldsymbol{T}\boldsymbol{T}^H + \rho\mathbf{I}_{M_2}\right)$. Nevertheless, it is still possible to verify that $\boldsymbol{T}^\dagger \triangleq \left(\boldsymbol{T}^H\boldsymbol{T}\right)^{-1}\boldsymbol{T}^H$ is also a Bezoutian matrix. Indeed, by applying the matrix inversion lemma to the expression $\rho\left(\boldsymbol{T}\boldsymbol{T}^H + \rho\mathbf{I}_{M_2}\right)^{-1}$ which appears in Equation (4.55), we have

$$\begin{aligned}
\rho\left(\boldsymbol{T}\boldsymbol{T}^H+\rho\mathbf{I}_{M_2}\right)^{-1} &= \rho\left[\frac{1}{\rho}\mathbf{I}_{M_2} - \frac{1}{\rho}\boldsymbol{T}\left(\frac{1}{\rho}\boldsymbol{T}^H\boldsymbol{T}+\mathbf{I}_{M_1}\right)^{-1}\boldsymbol{T}^H\frac{1}{\rho}\right] \\
&= \mathbf{I}_{M_2} - \boldsymbol{T}\left(\boldsymbol{T}^H\boldsymbol{T}+\rho\mathbf{I}_{M_1}\right)^{-1}\boldsymbol{T}^H \\
&= \mathbf{I}_{M_2} - \boldsymbol{T}\boldsymbol{T}_\rho^\dagger.
\end{aligned} \tag{4.56}$$

Now, by substituting Equation (4.56) into Equation (4.55), we end up with an expression which does not depend upon the matrix $\boldsymbol{T}\boldsymbol{T}^H$. In this case, one has

$$\begin{aligned}
\nabla_{\mathbf{Z}_\xi,\mathbf{Z}_\eta}\left(\boldsymbol{T}^\dagger\right) &= \lim_{\rho\to 0} \nabla_{\mathbf{Z}_\xi,\mathbf{Z}_\eta}\left(\boldsymbol{T}_\rho^\dagger\right) \\
&= -\boldsymbol{T}^\dagger \hat{\mathbf{P}} \hat{\mathbf{Q}}^H \boldsymbol{T}^\dagger + \left(\boldsymbol{T}^H\boldsymbol{T}\right)^{-1} \check{\mathbf{P}} \check{\mathbf{Q}}^H \left(\mathbf{I}_{M_2} - \boldsymbol{T}\boldsymbol{T}^\dagger\right) \\
&= \underbrace{\left[\left(\boldsymbol{T}^H\boldsymbol{T}\right)^{-1}\check{\mathbf{P}} \quad -\boldsymbol{T}^\dagger\hat{\mathbf{P}}\right]}_{\triangleq \mathbf{P}\in\mathbb{C}^{M_1\times 4}} \underbrace{\begin{bmatrix} \check{\mathbf{Q}}^H \left(\mathbf{I}_{M_2} - \boldsymbol{T}\boldsymbol{T}^\dagger\right) \\ \hat{\mathbf{Q}}^H \boldsymbol{T}^\dagger \end{bmatrix}}_{\triangleq \mathbf{Q}^H \in\mathbb{C}^{4\times M_2}} \\
&= \mathbf{P}\mathbf{Q}^H.
\end{aligned} \tag{4.57}$$

The reader should notice the similarity between the above (regularized) pseudo-inverses and the optimal ZF and MMSE linear solutions defined in Equations (4.16) and (4.17), apart from the transmitter-matrix inverse which appears only in the definitions of the optimal receiver matrices.[14] This means that the optimal receivers associated with ZP-ZJ transceivers are, essentially, Bezoutian matrices multiplied by a predefined unitary matrix $\overline{\mathbf{F}}$. Hence, inspired by the fact that the standard CP-OFDM and CP-SC-FD transceivers decompose the inverse of circulant channel matrices using DFT, IDFT, and diagonal matrices, it seems to be a good idea to look for decompositions of general Bezoutian matrices entailing the same type of matrices, namely DFT, IDFT, and diagonal.

4.4 DFT-BASED REPRESENTATIONS OF BEZOUTIAN MATRICES

Since the formal definitions and results in connection with the displacement operations are now available, we can proceed to investigate their use in the structured matrices related to reduced-redundancy transceivers. Thus, this section discusses the main mathematical results required to design block transceivers with reduced redundancy which are based on DFT.

From what we have described in Subsection 4.3.2, we know that Cauchy matrices are able to yield really simple Stein displacement matrices, as exemplified in Equation (4.44). This indicates that Cauchy matrices may have efficient representations, which could be exploited as long as we are able to transform Bezoutian matrices into Cauchy ones. In fact, this transformation is possible and many works have taken advantage of this possibility, as explained, for instance, in [27,28,49] and references therein. In other words, we are ultimately interested in transforming a Bezoutian-like matrix $\overline{\mathbf{G}}$ (see Equations (4.16) and (4.17)) into a new Cauchy-like matrix in order to benefit from the simplicity of Cauchy matrices. Figure 4.2 illustrates the strategy that we shall follow to decompose generalized (and possibly regularized) inverses of the effective channel matrix $\overline{\mathbf{H}}$ described in Equation (4.6). The entire process starts from a Toeplitz matrix and arrives at an efficient decomposition for the

[14] Remember that $\overline{\mathbf{H}}$ is a rectangular Toeplitz matrix (see Equation (4.6)).

generalized inverse of such a matrix. The crucial stages of this strategy are the efficient decompositions of Cauchy matrices and the efficient transformations of Bezoutian matrices into Cauchy matrices.

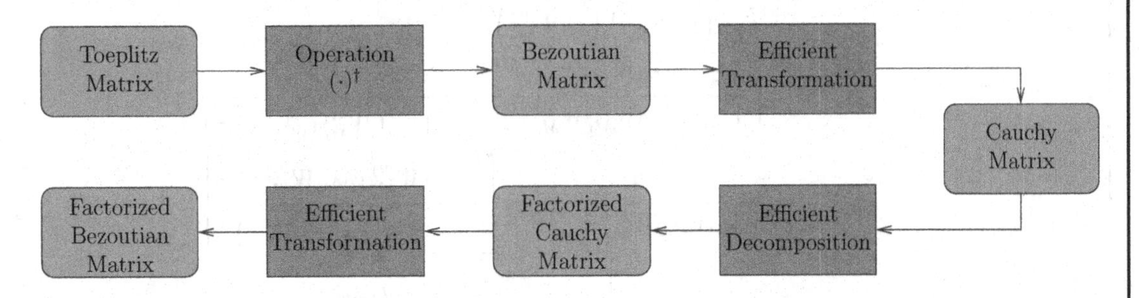

Figure 4.2: Strategy to decompose generalized inverses of Toeplitz matrices.

This section describes how one can decompose a particular type of Cauchy matrix using DFTs (Subsection 4.4.1) and, after that, how one can transform Bezoutian matrices into Cauchy matrices in a simple and effective manner (Subsection 4.4.2). As a result, one arrives at an efficient representation of Bezoutian matrices (Subsection 4.4.3).

4.4.1 REPRESENTATIONS OF CAUCHY MATRICES

The Cauchy matrices are known for their efficient decompositions. Let us consider a particular type of rectangular Cauchy matrix $\boldsymbol{C}_{\boldsymbol{\nu},\boldsymbol{\lambda}}$ with dimension $M_1 \times M_2$, in which the vectors $\boldsymbol{\nu} \in \mathbb{C}^{M_1 \times 1}$ and $\boldsymbol{\lambda} \in \mathbb{C}^{M_2 \times 1}$ are comprised of the distinct M_1th roots of $\nu \in \mathbb{C}$ and the M_2th roots of $\lambda \in \mathbb{C}$, respectively. That is,

$$\begin{aligned}
\boldsymbol{\nu} &\triangleq \begin{bmatrix} \nu_0 & \nu_1 & \cdots & \nu_{M_1-1} \end{bmatrix}^T, \\
\boldsymbol{\lambda} &\triangleq \begin{bmatrix} \lambda_0 & \lambda_1 & \cdots & \lambda_{M_2-1} \end{bmatrix}^T,
\end{aligned} \tag{4.58}$$

where $\nu_{m_1} \triangleq \nu_0 W_{M_1}^{m_1}$, for all m_1 within the set $\mathcal{M}_1$, whereas $\lambda_{m_2} \triangleq \lambda_0 W_{M_2}^{m_2}$, for all m_2 within the set $\mathcal{M}_2$ (see Equation (4.40)). Let us also assume that $M_2 \geq M_1$. Thus, from expression (4.43), we have

$$
\begin{aligned}
\left[\boldsymbol{C}_{\nu,\lambda}\right]_{m_1 m_2} &= \frac{1}{1-\nu_{m_1}\lambda_{m_2}} \\
&= \frac{1}{1-\left(\nu_0 W_{M_1}^{m_1}\right)\left(\lambda_0 W_{M_2}^{m_2}\right)} \\
&= \left[\frac{1-(\nu_0\lambda_0 W_{M_1}^{m_1} W_{M_2}^{m_2})^{M_1}}{1-(\nu_0\lambda_0 W_{M_1}^{m_1} W_{M_2}^{m_2})^{M_1}}\right] \times \frac{1}{1-\left(W_{M_1}^{m_1}\nu_0\lambda_0 W_{M_2}^{m_2}\right)} \\
&= \frac{1}{1-(\nu_0\lambda_0 W_{M_1}^{m_1} W_{M_2}^{m_2})^{M_1}} \left[\frac{1-(W_{M_1}^{m_1}\nu_0\lambda_0 W_{M_2}^{m_2})^{M_1}}{1-\left(W_{M_1}^{m_1}\nu_0\lambda_0 W_{M_2}^{m_2}\right)}\right] \\
&= \frac{1}{1-(\nu_0 W_{M_1}^{m_1})^{M_1}(\lambda_0 W_{M_2}^{m_2})^{M_1}} \sum_{m=0}^{M_1-1} \left(W_{M_1}^{m_1}\nu_0\lambda_0 W_{M_2}^{m_2}\right)^m \\
&= \frac{1}{1-\nu\lambda_{m_2}^{M_1}} \sum_{m=0}^{M_1-1} \left(W_{M_1}^{m_1 m}(\nu_0\lambda_0)^m\right)\left(W_{M_2}^{m m_2}\right),
\end{aligned} \tag{4.59}
$$

in which we considered that ν and λ are such that

$$
1-\nu\lambda_{m_2}^{M_1} \neq 0, \quad \forall m_2 \in \mathcal{M}_2. \tag{4.60}
$$

Now, let us observe that, for each m within the set $\mathcal{M}_1$, we have that $W_{M_2}^{m m_2}$ is the mth entry of the m_2th column of the matrix $\sqrt{M_2}\mathbf{W}_{M_2}$, while $W_{M_1}^{m_1 m}(\nu_0\lambda_0)^m$ is the mth entry of the m_1th row of the matrix $\sqrt{M_1}\mathbf{W}_{M_1}\left[\operatorname{diag}\{(\nu_0\lambda_0)^{m_1}\}_{m_1\in\mathcal{M}_1} \ \mathbf{0}_{M_1\times(M_2-M_1)}\right]$. Therefore, we can rewrite Equation (4.59) as

$$
\left[\boldsymbol{C}_{\nu,\lambda}\right]_{m_1 m_2} = \frac{\sqrt{M_2 M_1}}{1-\nu\lambda_{m_2}^{M_1}}\left[\mathbf{W}_{M_1}\left[\operatorname{diag}\{(\nu_0\lambda_0)^{m_1}\}_{m_1\in\mathcal{M}_1} \ \mathbf{0}_{M_1\times(M_2-M_1)}\right]\mathbf{W}_{M_2}\right]_{m_1 m_2}, \tag{4.61}
$$

yielding

$$
\boldsymbol{C}_{\nu,\lambda} = \mathbf{W}_{M_1}\left[\operatorname{diag}\{(\nu_0\lambda_0)^{m_1}\}_{m_1\in\mathcal{M}_1} \ \mathbf{0}_{M_1\times(M_2-M_1)}\right]\mathbf{W}_{M_2}\operatorname{diag}\left\{\frac{\sqrt{M_2 M_1}}{1-\nu\lambda_{m_2}^{M_1}}\right\}_{m_2\in\mathcal{M}_2}. \tag{4.62}
$$

Equation (4.62) contains an efficient matrix factorization which is also useful to express generalized Cauchy matrices. Indeed, as described in Subsection 4.3.2, we know that an $M_1 \times M_2$ generalized Cauchy matrix $\boldsymbol{C}_{\nu,\lambda}{}^{g}$ is such that

$$
\Delta_{\mathbf{D}_\nu,\mathbf{D}_\lambda}(\boldsymbol{C}_{\nu,\lambda}^{g}) = \sum_{r\in\mathcal{R}} \mathbf{p}_r \mathbf{q}_r^H, \tag{4.63}
$$

where $\mathcal{R} \triangleq \{0, 1, \cdots, R-1\}$, $\mathbf{p}_r \in \mathbb{C}^{M_1 \times 1}$ and $\mathbf{q}_r \in \mathbb{C}^{M_2 \times 1}$ are vectors, and (in fact, this is the main condition) $R \ll M_1 \leq M_2$. Observe that, by the definition of Stein displacement expressed in Equation (4.19), we have

$$\begin{aligned}
\left[\Delta_{\mathbf{D}_\nu, \mathbf{D}_\lambda}(\boldsymbol{C}^{\mathrm{g}}_{\nu,\lambda})\right]_{m_1 m_2} &= \left[\boldsymbol{C}^{\mathrm{g}}_{\nu,\lambda} - \mathbf{D}_\nu \boldsymbol{C}^{\mathrm{g}}_{\nu,\lambda} \mathbf{D}_\lambda\right]_{m_1 m_2} \\
&= \left[\boldsymbol{C}^{\mathrm{g}}_{\nu,\lambda}\right]_{m_1 m_2} - \left[\mathbf{D}_\nu \boldsymbol{C}^{\mathrm{g}}_{\nu,\lambda} \mathbf{D}_\lambda\right]_{m_1 m_2} \\
&= \left[\boldsymbol{C}^{\mathrm{g}}_{\nu,\lambda}\right]_{m_1 m_2} - \left[\mathbf{D}_\nu\right]_{m_1 m_1} \left[\boldsymbol{C}^{\mathrm{g}}_{\nu,\lambda}\right]_{m_1 m_2} \left[\mathbf{D}_\lambda\right]_{m_2 m_2} \\
&= (1 - \nu_{m_1} \lambda_{m_2}) \left[\boldsymbol{C}^{\mathrm{g}}_{\nu,\lambda}\right]_{m_1 m_2},
\end{aligned} \tag{4.64}$$

which, along with Equation (4.63), implies

$$\begin{aligned}
\left[\boldsymbol{C}^{\mathrm{g}}_{\nu,\lambda}\right]_{m_1 m_2} &= \frac{\left[\sum_{r \in \mathcal{R}} \mathbf{p}_r \mathbf{q}_r^H\right]_{m_1 m_2}}{(1 - \nu_{m_1} \lambda_{m_2})} \\
&= \frac{\sum_{r \in \mathcal{R}} p_{m_1,r} q^*_{m_2,r}}{(1 - \nu_{m_1} \lambda_{m_2})} \\
&= \sum_{r \in \mathcal{R}} p_{m_1,r} \frac{1}{(1 - \nu_{m_1} \lambda_{m_2})} q^*_{m_2,r} \\
&= \sum_{r \in \mathcal{R}} \left[\mathbf{D}_{\mathbf{p}_r}\right]_{m_1 m_1} \left[\boldsymbol{C}_{\nu,\lambda}\right]_{m_1 m_2} \left[\mathbf{D}^*_{\mathbf{q}_r}\right]_{m_2 m_2} \\
&= \left[\sum_{r \in \mathcal{R}} \mathbf{D}_{\mathbf{p}_r} \boldsymbol{C}_{\nu,\lambda} \mathbf{D}^*_{\mathbf{q}_r}\right]_{m_1 m_2},
\end{aligned} \tag{4.65}$$

in which, for each $r \in \mathcal{R}$, we define the following diagonal matrices:

$$\begin{aligned}
\mathbf{D}_{\mathbf{p}_r} &\triangleq \mathrm{diag}\{\mathbf{p}_r\} \in \mathbb{C}^{M_1 \times M_1}, \\
\mathbf{D}_{\mathbf{q}_r} &\triangleq \mathrm{diag}\{\mathbf{q}_r\} \in \mathbb{C}^{M_2 \times M_2}.
\end{aligned} \tag{4.66}$$

Therefore, based on Equations (4.62) and (4.65), we conclude that the generalized Cauchy matrix $\boldsymbol{C}^{\mathrm{g}}_{\nu,\lambda}$ admits the following efficient decomposition:

$$\begin{aligned}
\boldsymbol{C}^{\mathrm{g}}_{\nu,\lambda} = \sum_{r \in \mathcal{R}} \mathbf{D}_{\mathbf{p}_r} \mathbf{W}_{M_1} &\left[\, \mathrm{diag}\{(\nu_0 \lambda_0)^{m_1}\}_{m_1 \in \mathcal{M}_1} \;\; \mathbf{0}_{M_1 \times (M_2 - M_1)} \right] \\
&\times \mathbf{W}_{M_2} \mathbf{D}^*_{\mathbf{q}_r} \mathrm{diag}\left\{\frac{\sqrt{M_2 M_1}}{1 - \nu \lambda^{M_1}_{m_2}}\right\}_{m_2 \in \mathcal{M}_2}.
\end{aligned} \tag{4.67}$$

Equation (4.67) is a quite efficient way of decomposing a given generalized Cauchy matrix, since the decomposition only employs DFT and diagonal matrices. What remains is to find a way of transforming a given rectangular Bezoutian matrix into such a type of Cauchy matrix.

4.4.2 TRANSFORMATIONS OF BEZOUTIAN MATRICES INTO CAUCHY MATRICES

This subsection describes a computationally efficient way to transform rectangular Bezoutian matrices into rectangular Cauchy matrices in order to benefit from the fact that Cauchy matrices enjoy efficient decompositions. The transformation which we shall describe is based on the displacement operators of Sylvester and Stein.[15]

From a displacement-approach point of view, we know that Bezoutian and Cauchy matrices have an inherent feature in common: their resulting displacement matrices have small rank, as compared to the dimensions of the original matrices. At a first look, it seems that the similarities end here, since Bezoutian matrices employ Sylvester displacements, whereas Cauchy matrices use Stein displacements. In addition, the operator matrices utilized to compress both classes of matrices are also of different types. These facts might erroneously lead us to conclude that the transformation from Bezoutian to Cauchy matrices is not possible.

Fortunately, this is not the case, i.e., it is possible to transform Bezoutian into Cauchy matrices. Indeed, the difference between the Sylvester and Stein displacements is not a problem since we know how to convert one into another, as explained in Equations (4.46) and (4.47). With regard to the difference between the operator-matrix types, a Bezoutian matrix uses ν-circulant operator matrices, such as $\mathbf{Z}_\nu$, whereas a Cauchy matrix uses diagonal operator matrices, let us say $\mathbf{D}_{\boldsymbol{\nu}}$. Even though these operator matrices are quite different, there is a simple relationship between them which will help us link Bezoutian to Cauchy matrices.

Consider the particular case in which vector $\boldsymbol{\nu}$ is comprised of all distinct Mth roots of $\nu \in \mathbb{C}$, i.e.,

$$\boldsymbol{\nu} \triangleq \begin{bmatrix} \nu_0 & \nu_1 & \cdots & \nu_{M-1} \end{bmatrix}^T, \tag{4.68}$$

where $\nu_0 = |\nu|^{\frac{1}{M}} e^{j\frac{\angle \nu}{M}}$ and $\nu_m = \nu_0 W_M^m$, for all m within the set $\mathcal{M}$. Now, by remembering the effect of left-multiplying a Vandermonde matrix $\boldsymbol{V}_{\boldsymbol{\nu}}$ by a diagonal matrix $\mathbf{D}_{\boldsymbol{\nu}}$, as described in Equation (4.39) (recall that $\nu_m^M = \nu$, for any m in $\mathcal{M}$), we see that the resulting matrix $\mathbf{D}_{\boldsymbol{\nu}}\boldsymbol{V}_{\boldsymbol{\nu}}$ is equal to the matrix $\boldsymbol{V}_{\boldsymbol{\nu}}\mathbf{Z}_\nu$, thus relating the diagonal operator matrix to the ν-circulant operator matrix through the use of a Vandermonde matrix.

Such a fact can also be derived in a more formal way as follows: first consider that $m_1 \in \mathcal{M} \setminus \{M-1\}$. Thus, $\left[\mathbf{D}_{\boldsymbol{\nu}}\boldsymbol{V}_{\boldsymbol{\nu}}\right]_{m_2 m_1} = \nu_{m_2} W_M^{m_2 m_1} \nu_0^{m_1} = \nu_0 W_M^{m_2} W_M^{m_2 m_1} \nu_0^{m_1} = W_M^{m_2(m_1+1)} \nu_0^{m_1+1} = [\boldsymbol{V}_{\boldsymbol{\nu}}]_{m_2(m_1+1)}$. Now consider that $m_1 = M-1$. In this case, we have that $\left[\mathbf{D}_{\boldsymbol{\nu}}\boldsymbol{V}_{\boldsymbol{\nu}}\right]_{m_2(M-1)} = \nu_{m_2} W_M^{m_2(M-1)} \nu_0^{M-1} = W_M^{m_2 M} \nu_0^M = \nu = \nu\, [\boldsymbol{V}_{\boldsymbol{\nu}}]_{m_2 0}$. We can therefore conclude that

$$\boldsymbol{V}_{\boldsymbol{\nu}}\mathbf{Z}_\nu = \mathbf{D}_{\boldsymbol{\nu}}\boldsymbol{V}_{\boldsymbol{\nu}} \tag{4.69}$$

[15] The results derived here are heavily based on the outstanding works developed by Heinig and Rost, for example, in [27]. These authors, however, employed a polynomial formulation to develop most of their results.

and, for the useful case where $\nu \neq 0$, one has

$$\mathbf{Z}_\nu = \boldsymbol{V}_\nu^{-1}\mathbf{D}_\nu \boldsymbol{V}_\nu. \tag{4.70}$$

We can now proceed to derive the expected transformation from Bezoutian to Cauchy matrices. We know that, given two natural numbers M_1 and M_2, with $M_1 \leq M_2$, and given two non-zero complex numbers ξ and η, an $M_1 \times M_2$ complex-valued Bezoutian matrix $\boldsymbol{B}$ is such that $\nabla_{\mathbf{Z}_\xi,\mathbf{Z}_\eta}(\boldsymbol{B}) = \mathbf{P}\mathbf{Q}^H$, where the operator matrices have compatible dimensions and the resulting displacement rank is much smaller than M_1. Our first task is to change from the Sylvester to the Stein displacement and, after that, to convert the circulant operator matrices into diagonal ones. Thus, by defining $\overline{\eta} \triangleq 1/\eta$ and by using the relations in Equations (4.22), (4.47), and (4.70), we have

$$\begin{aligned}
\mathbf{P}\mathbf{Q}^H &= \nabla_{\mathbf{Z}_\xi,\mathbf{Z}_\eta}(\boldsymbol{B}) \\
&= \mathbf{Z}_\xi \boldsymbol{B} - \boldsymbol{B}\mathbf{Z}_\eta \\
&= -\Delta_{\mathbf{Z}_\xi,\mathbf{Z}_{1/\eta}^T}(\boldsymbol{B})\mathbf{Z}_\eta \\
&= -\left(\boldsymbol{B} - \mathbf{Z}_\xi \boldsymbol{B}\mathbf{Z}_{\overline{\eta}}^T\right)\mathbf{Z}_\eta \\
&= -\left(\boldsymbol{B} - \boldsymbol{V}_\xi^{-1}\mathbf{D}_\xi \boldsymbol{V}_\xi \boldsymbol{B}\boldsymbol{V}_{\overline{\eta}}^T \mathbf{D}_{\overline{\eta}} \boldsymbol{V}_{\overline{\eta}}^{-T}\right)\mathbf{Z}_\eta \\
&= -\boldsymbol{V}_\xi^{-1}\left[\left(\boldsymbol{V}_\xi \boldsymbol{B}\boldsymbol{V}_{\overline{\eta}}^T\right) - \mathbf{D}_\xi\left(\boldsymbol{V}_\xi \boldsymbol{B}\boldsymbol{V}_{\overline{\eta}}^T\right)\mathbf{D}_{\overline{\eta}}\right]\boldsymbol{V}_{\overline{\eta}}^{-T}\mathbf{Z}_\eta \\
&= -\boldsymbol{V}_\xi^{-1}\Delta_{\mathbf{D}_\xi,\mathbf{D}_{\overline{\eta}}}(\boldsymbol{V}_\xi \boldsymbol{B}\boldsymbol{V}_{\overline{\eta}}^T)\boldsymbol{V}_{\overline{\eta}}^{-T}\mathbf{Z}_\eta,
\end{aligned} \tag{4.71}$$

implying that

$$\begin{aligned}
\Delta_{\mathbf{D}_\xi,\mathbf{D}_{\overline{\eta}}}(\boldsymbol{V}_\xi \boldsymbol{B}\boldsymbol{V}_{\overline{\eta}}^T) &= -\boldsymbol{V}_\xi \mathbf{P}\mathbf{Q}^H \mathbf{Z}_{\overline{\eta}^*}^H \boldsymbol{V}_{\overline{\eta}}^T \\
&= \underbrace{(-\boldsymbol{V}_\xi \mathbf{P})}_{\triangleq \mathbf{P}'}\underbrace{\left(\boldsymbol{V}_{\overline{\eta}}^* \mathbf{Z}_{\overline{\eta}^*}\mathbf{Q}\right)^H}_{\triangleq \mathbf{Q}'^H} \\
&= \mathbf{P}'\mathbf{Q}'^H \\
&= \begin{bmatrix}\mathbf{p}_0' & \mathbf{p}_1' & \cdots & \mathbf{p}_{R-1}'\end{bmatrix}\begin{bmatrix}\mathbf{q}_0' & \mathbf{q}_1' & \cdots & \mathbf{q}_{R-1}'\end{bmatrix}^H,
\end{aligned} \tag{4.72}$$

in which R is a natural number such that $M_1 \gg R \geq \text{rank}\{\nabla_{\mathbf{Z}_\xi,\mathbf{Z}_\eta}(\boldsymbol{B})\}$.

Equation (4.72) means that matrix $\boldsymbol{V}_\xi \boldsymbol{B}\boldsymbol{V}_{\overline{\eta}}^T$ is a generalized Cauchy matrix, since its related Stein displacement matrix is such that

$$\text{rank}\{\Delta_{\mathbf{D}_\xi,\mathbf{D}_{\overline{\eta}}}(\boldsymbol{V}_\xi \boldsymbol{B}\boldsymbol{V}_{\overline{\eta}}^T)\} = \text{rank}\{\nabla_{\mathbf{Z}_\xi,\mathbf{Z}_\eta}(\boldsymbol{B})\} \leq R \ll M_1. \tag{4.73}$$

4.4.3 EFFICIENT BEZOUTIAN DECOMPOSITIONS

We are now in a position to describe how to decompose a given rectangular Bezoutian matrix by employing computationally efficient matrices like DFT, IDFT, and diagonal matrices. Indeed, given the expression in Equation (4.72), we can apply the result to Equation (4.67), replacing ν by ξ, λ by $\overline{\eta} = 1/\eta$, $\mathbf{p}_r$ by $\mathbf{p}'_r$, $\mathbf{q}_r$ by $\mathbf{q}'_r$, and $\boldsymbol{C}^{\mathrm{g}}_{\nu,\lambda}$ by $\boldsymbol{V}_{\xi}\boldsymbol{B}\boldsymbol{V}^T_{\overline{\eta}}$. The resulting expression is

$$\begin{aligned}\boldsymbol{V}_{\xi}\boldsymbol{B}\boldsymbol{V}^T_{\overline{\eta}} =& \sum_{r\in\mathcal{R}} \mathbf{D}_{\mathbf{p}'_r}\mathbf{W}_{M_1}\left[\operatorname{diag}\{(\xi_0\overline{\eta}_0)^{m_1}\}_{m_1\in\mathcal{M}_1} \quad \mathbf{0}_{M_1\times(M_2-M_1)}\right] \\ &\times \mathbf{W}_{M_2}\mathbf{D}^*_{\mathbf{q}'_r}\operatorname{diag}\left\{\frac{\sqrt{M_2M_1}}{1-\xi\overline{\eta}^{-M_1}_{m_2}}\right\}_{m_2\in\mathcal{M}_2},\end{aligned} \tag{4.74}$$

which yields

$$\begin{aligned}\boldsymbol{B} =& \boldsymbol{V}_{\xi}^{-1}\Bigg[\sum_{r\in\mathcal{R}} \mathbf{D}_{\mathbf{p}'_r}\mathbf{W}_{M_1}\left[\operatorname{diag}\{(\xi_0\overline{\eta}_0)^{m_1}\}_{m_1\in\mathcal{M}_1} \quad \mathbf{0}_{M_1\times(M_2-M_1)}\right] \\ &\times \mathbf{W}_{M_2}\mathbf{D}^*_{\mathbf{q}'_r}\operatorname{diag}\left\{\frac{\sqrt{M_2M_1}}{1-\xi\overline{\eta}^{-M_1}_{m_2}}\right\}_{m_2\in\mathcal{M}_2}\Bigg]\boldsymbol{V}^{-T}_{\overline{\eta}}.\end{aligned} \tag{4.75}$$

Equation (4.75) represents the last step of the displacement approach, namely the decompression operation. It is worth mentioning that such expression is a rather efficient decomposition of a general rectangular Bezoutian matrix, since it employs only DFT and diagonal matrices, and considering that a given Vandermonde matrix can be described as a product of a DFT and a diagonal matrix, as described in Equation (4.42).

Example 4.3 (Bezoutian Factorization) Let us analyze a numerical toy example which will help us better understand the computations involved in Bezoutian factorizations. Consider the following 4×3 Toeplitz matrix

$$\boldsymbol{T} = \begin{bmatrix} 2 & 1 & 0 \\ 3 & 2 & 1 \\ 0 & 3 & 2 \\ 0 & 0 & 3 \end{bmatrix}, \tag{4.76}$$

whose associated Moore-Penrose pseudo-inverse $\boldsymbol{T}^{\dagger} = \left(\boldsymbol{T}^H\boldsymbol{T}\right)^{-1}\boldsymbol{T}^H$ is given by[16]

$$\boldsymbol{T}^{\dagger} \approx \begin{bmatrix} 0.1633 & 0.2245 & -0.2041 & 0.0612 \\ -0.0028 & 0.0019 & 0.3330 & -0.2226 \\ -0.0334 & 0.0223 & -0.0037 & 0.3284 \end{bmatrix}. \tag{4.77}$$

As discussed in Subsection 4.3.3, matrix $\boldsymbol{T}^{\dagger}$ is a particular type of Bezoutian matrix. We could therefore apply the result described in Equation (4.75), with $M_1 = 3$ and $M_2 = 4$.[17] Assume, for

[16]The symbol "$\approx$" is employed rather than "$=$" since the entries of $\boldsymbol{T}^{\dagger}$ are in fact real numbers with more than four decimal digits of precision.

[17]Note that neither M_1 nor M_2 is much larger than the displacement rank of $\boldsymbol{T}^{\dagger}$. Nevertheless, the obtained factorization does not depend on having $R \ll \min\{M_1, M_2\}$. Thus, even though it is a bit misleading to say that $\boldsymbol{T}^{\dagger}$ is a Bezoutian matrix due to its small dimensions, we still can apply the result in Equation (4.75).

instance, that $\xi = 2$ and $\eta = 1 = \overline{\eta}$. With such assumptions, most of the matrices appearing in the decomposition in Equation (4.75) can be readily defined, where $\xi_0 = \sqrt[3]{2}$ (note that $\angle\xi = 0$), $\overline{\eta}_0 = \sqrt[4]{1} = 1$ ($\angle\overline{\eta} = 0$ as well), and $\overline{\eta}_{m_2}^{M_1} = (\overline{\eta}_0 W_{M_2}^{m_2})^{M_1} = (\mathrm{e}^{-\mathrm{j}\frac{2\pi}{4}m_2})^3 = \mathrm{e}^{-\mathrm{j}\frac{3\pi}{2}m_2} = \mathrm{j}^{m_2}$, for each $m_2 \in \{0, 1, 2, 3\}$. Thus, by using Equation (4.42), we have

$$
\begin{aligned}
\boldsymbol{V}_{\xi}^{-1} &= \left(\sqrt{3}\mathbf{W}_3 \mathrm{diag}\{2^{\frac{m_1}{3}}\}_{m_1=0}^{m_1=2}\right)^{-1} \\
&= \frac{1}{\sqrt{3}}\left(\mathrm{diag}\{2^{-\frac{m_1}{3}}\}_{m_1=0}^{m_1=2}\right)\mathbf{W}_3^H \\
&= \frac{1}{3}\begin{bmatrix} 1 & 0 & 0 \\ 0 & 2^{-\frac{1}{3}} & 0 \\ 0 & 0 & 4^{-\frac{1}{3}} \end{bmatrix}\begin{bmatrix} 1 & 1 & 1 \\ 1 & \mathrm{e}^{\mathrm{j}\frac{2\pi}{3}} & \mathrm{e}^{\mathrm{j}\frac{4\pi}{3}} \\ 1 & \mathrm{e}^{\mathrm{j}\frac{4\pi}{3}} & \mathrm{e}^{\mathrm{j}\frac{2\pi}{3}} \end{bmatrix},
\end{aligned} \tag{4.78}
$$

$$
\begin{aligned}
\boldsymbol{V}_{\overline{\eta}}^{-T} &= \left(\sqrt{4}\mathbf{W}_4 \mathrm{diag}\{1^{\frac{m_2}{4}}\}_{m_2=0}^{m_2=3}\right)^{-T} \\
&= \frac{1}{2}\mathbf{W}_4^H \\
&= \frac{1}{4}\begin{bmatrix} 1 & 1 & 1 & 1 \\ 1 & \mathrm{j} & -1 & -\mathrm{j} \\ 1 & -1 & 1 & -1 \\ 1 & -\mathrm{j} & -1 & \mathrm{j} \end{bmatrix}.
\end{aligned} \tag{4.79}
$$

In addition, we have

$$
\left[\, \mathrm{diag}\{(\xi_0\overline{\eta}_0)^{m_1}\}_{m_1\in\mathcal{M}_1} \;\; \mathbf{0}_{M_1\times(M_2-M_1)} \,\right] = \begin{bmatrix} 1 & 0 & 0 & 0 \\ 0 & 2^{\frac{1}{3}} & 0 & 0 \\ 0 & 0 & 4^{\frac{1}{3}} & 0 \end{bmatrix}, \tag{4.80}
$$

and

$$
\mathrm{diag}\left\{\frac{\sqrt{M_2 M_1}}{1 - \xi\overline{\eta}_{m_2}^{M_1}}\right\}_{m_2\in\mathcal{M}_2} = 2\sqrt{3}\begin{bmatrix} -1 & 0 & 0 & 0 \\ 0 & \frac{1}{1-2\mathrm{j}} & 0 & 0 \\ 0 & 0 & \frac{1}{3} & 0 \\ 0 & 0 & 0 & \frac{1}{1+2\mathrm{j}} \end{bmatrix}. \tag{4.81}
$$

The reader should note that we have just defined almost all matrices to be employed in the factorization, but we have not used any information about matrix $\boldsymbol{T}$ yet. Indeed, all information of $\boldsymbol{T}$ is contained in the entries of the diagonal matrices $\mathbf{D}_{\mathbf{p}'_r}$ and $\mathbf{D}_{\mathbf{q}'_r}$, with $r \in \{0, 1, \cdots, R-1\}$. Thus, we have to determine vectors $\mathbf{p}'_r$ and $\mathbf{q}'_r$ based on Equation (4.72). But, before that, we have to determine matrices $\mathbf{P}$ and $\mathbf{Q}$ based on Equation (4.57). These latter matrices, on the other hand, depend on the knowledge of the displacement matrices $\nabla_{\mathbf{Z}_\eta,\mathbf{Z}_\xi}(\boldsymbol{T}) = \hat{\mathbf{P}}\hat{\mathbf{Q}}^H$ and $\nabla_{\mathbf{Z}_\xi,\mathbf{Z}_\eta}(\boldsymbol{T}^H) = \check{\mathbf{P}}\check{\mathbf{Q}}^H$, which can be straightforwardly acquired using Equation (4.37), as follows:

$$\hat{\mathbf{P}} = \begin{bmatrix} 1 & -4 \\ 0 & -6 \\ 0 & 1 \\ 0 & 2 \end{bmatrix}, \tag{4.82}$$

$$\hat{\mathbf{Q}} = \begin{bmatrix} -1 & 0 \\ 0 & 0 \\ 3 & 1 \end{bmatrix}, \tag{4.83}$$

and

$$\check{\mathbf{P}} = \begin{bmatrix} 1 & -2 \\ 0 & -1 \\ 0 & 0 \end{bmatrix}, \tag{4.84}$$

$$\check{\mathbf{Q}} = \begin{bmatrix} -3 & 0 \\ 2 & 0 \\ 4 & 0 \\ 6 & 1 \end{bmatrix}. \tag{4.85}$$

Now, Equation (4.57) gives us

$$\mathbf{P} \approx \begin{bmatrix} 0.1224 & -0.1633 & -0.1633 & 2.0816 \\ -0.0816 & 0.0028 & 0.0028 & 0.1122 \\ 0.0204 & 0.0334 & 0.0334 & -0.6531 \end{bmatrix}, \tag{4.86}$$

$$\mathbf{Q} \approx \begin{bmatrix} -2.0288 & 0.1002 & -0.2635 & -0.0334 \\ 1.3525 & -0.0668 & -0.1577 & 0.0223 \\ -0.2254 & 0.0111 & 0.1929 & -0.0037 \\ -0.3006 & 0.0148 & 0.9239 & 0.3284 \end{bmatrix}. \tag{4.87}$$

Thus, using Equation (4.72), we have $R = 4$ and

$$\mathbf{P}' \approx \begin{bmatrix} -0.0520 & 0.1067 & 0.1067 & -1.1864 \\ -0.1577 - 0.1171\mathrm{j} & 0.1915 - 0.0429\mathrm{j} & 0.1915 - 0.0429\mathrm{j} & -2.5293 + 1.0203\mathrm{j} \\ -0.1577 + 0.1171\mathrm{j} & 0.1915 + 0.0429\mathrm{j} & 0.1915 + 0.0429\mathrm{j} & -2.5293 - 1.0203\mathrm{j} \end{bmatrix}, \tag{4.88}$$

$$\mathbf{Q}' \approx \begin{bmatrix} -1.2022 & 0.0594 & 0.6957 & 0.3135 \\ -1.6531 - 1.8033\mathrm{j} & 0.0816 + 0.0891\mathrm{j} & 1.0816 - 0.4564\mathrm{j} & 0.3061 - 0.0297\mathrm{j} \\ 3.3061 & -0.1633 & 0.8367 & 0.3878 \\ -1.6531 + 1.8033\mathrm{j} & 0.0816 - 0.0891\mathrm{j} & 1.0816 + 0.4564\mathrm{j} & 0.3061 + 0.0297\mathrm{j} \end{bmatrix}. \tag{4.89}$$

By using matrices $\mathbf{P}'$ and $\mathbf{Q}'$ we can define the diagonal matrices $\mathbf{D}_{\mathbf{p}'_r}$ and $\mathbf{D}_{\mathbf{q}'_r}$, with $r \in \{0, 1, 2, 3\}$. For example, we have

$$\mathbf{D}_{\mathbf{q}'_1} \approx \begin{bmatrix} 0.0594 & 0 & 0 & 0 \\ 0 & 0.0816 + 0.0891\mathrm{j} & 0 & 0 \\ 0 & 0 & -0.1633 & 0 \\ 0 & 0 & 0 & 0.0816 - 0.0891\mathrm{j} \end{bmatrix}. \tag{4.90}$$

It is possible to verify that Equation (4.75) holds for this particular case using the numerical values in Equations (4.78), (4.79), (4.80), (4.81), (4.88), and (4.89). □

4.5 REDUCED-REDUNDANCY SYSTEMS

This section addresses the issue of designing memoryless LTI transceivers with reduced redundancy which are amenable to superfast implementations at the transmitter and receiver ends. The transceivers are generated by exploiting the displacement structure of the Toeplitz equivalent channel matrix, thus yielding systems whose corresponding asymptotic computational complexity is comparable to the complexity of the conventional OFDM and SC-FD transceivers with respect to the equalization process. The reduced-redundancy transceivers may lead to higher data throughputs when compared to OFDM and SC-FD systems, due to the reduced amount of redundancy which is employed.

As pointed out before, the equivalent channel matrix $\overline{\mathbf{H}}$ associated with reduced-redundancy ZP-ZJ systems is a rectangular Toeplitz matrix (see Equation (4.6)). In addition, Equations (4.55) and (4.57) show that any (regularized) Moore-Penrose pseudo-inverse of a given full-rank Toeplitz matrix is indeed a Bezoutian matrix. Moreover, if we define the *transmitter-independent receiver matrix* $\mathbf{K}$ as

$$\mathbf{K} \triangleq \overline{\mathbf{F}}\,\overline{\mathbf{G}} \in \mathbb{C}^{M \times (M+2K-L)}, \tag{4.91}$$

with $2K \geq L$, then one can easily verify based on Equations (4.16) and (4.17) that

$$\mathbf{K}_{\text{ZF}} = \overline{\mathbf{H}}^{\dagger}, \tag{4.92}$$

$$\mathbf{K}_{\text{MMSE}} = \left(\overline{\mathbf{H}}^H \overline{\mathbf{H}} + \frac{\sigma_v^2}{\sigma_s^2}\mathbf{I}_M\right)^{-1} \overline{\mathbf{H}}^H \tag{4.93}$$

are (possibly regularized) pseudo-inverses of the tall channel Toeplitz matrix $\overline{\mathbf{H}}$, thus implying that the transmitter-independent receiver matrix is also a particular type of rectangular Bezoutian matrix.

Hence, we can tailor the efficient decompositions of Bezoutian matrices previously described in Subsection 4.4.3 to the particular cases of MMSE and ZF transmitter-independent receiver matrices. As a result, a family of superfast multicarrier and single-carrier linear transceivers are derived with their respective structures.

Firstly, we must carefully choose both parameters ξ and η which appear in the definition of the operator matrices employed in the compression stage of both the Toeplitz channel matrix and the receiver Bezoutian matrix. Let us assume that $\xi = 1$ and $\eta = \mathrm{e}^{\mathrm{j}\frac{\pi}{M}}$. In this case, by using Equation (4.37) one can readily compute the displacement-generator pair of matrices $(\hat{\mathbf{P}}, \hat{\mathbf{Q}}) \in \mathbb{C}^{(M+2K-L)\times 2} \times \mathbb{C}^{M\times 2}$ and $(\check{\mathbf{P}}, \check{\mathbf{Q}}) \in \mathbb{C}^{M\times 2} \times \mathbb{C}^{(M+2K-L)\times 2}$ associated with matrices $\overline{\mathbf{H}}$ and $\overline{\mathbf{H}}^H$, respectively, where $\nabla_{\mathbf{Z}_\eta,\mathbf{Z}_\xi}\left(\overline{\mathbf{H}}\right) = \hat{\mathbf{P}}\hat{\mathbf{Q}}^H$ and $\nabla_{\mathbf{Z}_\xi,\mathbf{Z}_\eta}\left(\overline{\mathbf{H}}^H\right) = \check{\mathbf{P}}\check{\mathbf{Q}}^H$. By using such displacement-generator pairs, one can precompute the displacement-generator pair $(\mathbf{P}, \mathbf{Q}) \in \mathbb{C}^{M\times 4} \times \mathbb{C}^{(M+2K-L)\times 4}$ associated with the transmitter-independent receiver matrix, where $\nabla_{\mathbf{Z}_\xi,\mathbf{Z}_\eta}(\mathbf{K}) = \mathbf{P}\mathbf{Q}^H$. Indeed, for the MMSE solution, Equation (4.55) gives us

$$\mathbf{P} = \left[\left(\overline{\mathbf{H}}^H\overline{\mathbf{H}} + \frac{\sigma_v^2}{\sigma_s^2}\mathbf{I}_M\right)^{-1}\check{\mathbf{P}} \quad -\mathbf{K}_{\mathrm{MMSE}}\hat{\mathbf{P}}\right], \tag{4.94}$$

$$\mathbf{Q} = \left[\frac{\sigma_v^2}{\sigma_s^2}\left(\overline{\mathbf{H}}\,\overline{\mathbf{H}}^H + \frac{\sigma_v^2}{\sigma_s^2}\mathbf{I}_{(M+2K-L)}\right)^{-H}\check{\mathbf{Q}} \quad \mathbf{K}_{\mathrm{MMSE}}^H\hat{\mathbf{Q}}\right], \tag{4.95}$$

whereas, for the ZF solution, Equation (4.57) gives us

$$\mathbf{P} = \left[\left(\overline{\mathbf{H}}^H\overline{\mathbf{H}}\right)^{-1}\check{\mathbf{P}} \quad -\mathbf{K}_{\mathrm{ZF}}\hat{\mathbf{P}}\right], \tag{4.96}$$

$$\mathbf{Q} = \left[\left(\mathbf{I}_{(M+2K-L)} - \overline{\mathbf{H}}\mathbf{K}_{\mathrm{ZF}}\right)^H\check{\mathbf{Q}} \quad \mathbf{K}_{\mathrm{ZF}}^H\hat{\mathbf{Q}}\right]. \tag{4.97}$$

Thus, by considering the previous compressed forms of the Bezoutian matrix $\mathbf{K}$ for either the MMSE or the ZF solutions, one can employ the factorization in Equation (4.75), which depends on the Vandermonde matrices $\mathbf{V}_\xi$ and $\mathbf{V}_{\overline{\eta}}$. Since $\xi = 1$ and $\overline{\eta} = 1/\eta = \mathrm{e}^{-\mathrm{j}\frac{\pi}{M}}$, then

$$\begin{aligned}\xi_0 &= |\xi|^{1/M}\mathrm{e}^{\mathrm{j}\frac{\angle\xi}{M}}\\ &= 1\times \mathrm{e}^0\\ &= 1\end{aligned} \tag{4.98}$$

and

$$\begin{aligned}\overline{\eta}_0 &= |\overline{\eta}|^{1/(M+2K-L)}\mathrm{e}^{\mathrm{j}\frac{\angle\overline{\eta}}{(M+2K-L)}}\\ &= 1\times \mathrm{e}^{\frac{-\mathrm{j}\pi}{M(M+2K-L)}}\\ &= \mathrm{e}^{\frac{-\mathrm{j}\pi}{M(M+2K-L)}}.\end{aligned} \tag{4.99}$$

These facts, along with Equation (4.42), imply that

$$\mathbf{V}_\xi = \sqrt{M}\mathbf{W}_M, \tag{4.100}$$

whereas

$$\mathbf{V}_{\overline{\eta}} = \sqrt{M+2K-L}\times\mathbf{W}_{(M+2K-L)}\mathrm{diag}\{\mathrm{e}^{\frac{-\mathrm{j}\pi m}{M(M+2K-L)}}\}_{m=0}^{(M+2K-L-1)}. \tag{4.101}$$

We can therefore apply the decomposition presented in Equation (4.75) to obtain the following result:

$$\mathbf{K} = \mathbf{W}_M^H \left[\sum_{r=0}^{3} \mathbf{D}_{\overline{\mathbf{p}}_r} \mathbf{W}_M \left[\mathbf{D}_M \quad \mathbf{0}_{M\times(2K-L)} \right] \mathbf{W}_{(M+2K-L)} \mathbf{D}_{\overline{\mathbf{q}}_r}^* \right] \mathbf{W}_{(M+2K-L)}^H \mathbf{D}_{(M+2K-L)}^H, \tag{4.102}$$

in which the diagonal matrices $\mathbf{D}_M$ and $\mathbf{D}_{(M+2K-L)}$ do not depend on $\mathbf{K}$ and are, respectively, defined as

$$\mathbf{D}_M \triangleq \operatorname{diag}\left\{ \mathrm{e}^{\frac{-\mathrm{j}\pi m}{M(M+2K-L)}} \right\}_{m=0}^{M-1}, \tag{4.103}$$

$$\mathbf{D}_{(M+2K-L)} \triangleq \operatorname{diag}\left\{ \mathrm{e}^{\frac{-\mathrm{j}\pi m}{M(M+2K-L)}} \right\}_{m=0}^{M+2K-L-1}. \tag{4.104}$$

In addition, based on Equations (4.72) and (4.75), the pair of matrices $(\overline{\mathbf{P}}, \overline{\mathbf{Q}}) \in \mathbb{C}^{M\times 4} \times \mathbb{C}^{(M+2K-L)\times 4}$ are defined as

$$\begin{aligned} \overline{\mathbf{P}} &\triangleq [\overline{\mathbf{p}}_0 \quad \overline{\mathbf{p}}_1 \quad \overline{\mathbf{p}}_2 \quad \overline{\mathbf{p}}_3] \\ &= -\boldsymbol{V}_{\xi}\mathbf{P}, \end{aligned} \tag{4.105}$$

$$\begin{aligned} \overline{\mathbf{Q}} &\triangleq [\overline{\mathbf{q}}_0 \quad \overline{\mathbf{q}}_1 \quad \overline{\mathbf{q}}_2 \quad \overline{\mathbf{q}}_3] \\ &= \left(\operatorname{diag}\left\{ \frac{1}{1-\xi^*(\overline{\eta}_m^*)^M} \right\}_{m=0}^{M+2K-L-1} \right) \mathbf{V}_{\overline{\eta}}^* \mathbf{Z}_{\overline{\eta}^*} \mathbf{Q}, \end{aligned} \tag{4.106}$$

where $\mathbf{P}$ and $\mathbf{Q}$ are defined in Equations (4.94) and (4.95) for the MMSE solution, and in Equations (4.96) and (4.97) for the ZF solution, whereas $\xi = 1$ and $\overline{\eta} = \mathrm{e}^{-\mathrm{j}\frac{\pi}{M}}$.

Note that the choices of ξ and η were quite arbitrary. We have chosen $\xi = 1$, since, in the case of multicarrier systems, we would like to cancel out the last IDFT matrix appearing in the decomposition of matrix $\mathbf{K}$ described in Equation (4.102). Indeed, in the multicarrier systems, the receiver matrix is $\overline{\mathbf{G}} = \mathbf{W}_M\mathbf{K}$. If $\xi \neq 1$, one would not be able to cancel out the DFT matrix with the last IDFT matrix presented in the decomposition of $\mathbf{K}$, due to the presence of an additional diagonal matrix. After fixing $\xi = 1$, we have chosen η in such a way that $1 - \xi\overline{\eta}_m^M = 1 - \overline{\eta}_m^M \neq 0$, for all m within the set $\{0, 1, \cdots, M+2K-L-1\}$, where $\overline{\eta} = 1/\eta$. There are infinite possible choices for η and we have arbitrarily chosen $\eta = \mathrm{e}^{\mathrm{j}\frac{\pi}{M}}$. Note that, for this choice of η, one has

$$\begin{aligned} \overline{\eta}_m^M &= (\overline{\eta}_0 W_{(M+2K-L)}^m)^M \\ &= \mathrm{e}^{\frac{-\mathrm{j}\pi}{(M+2K-L)}} \mathrm{e}^{\frac{-\mathrm{j}2\pi m M}{(M+2K-L)}} \\ &= \mathrm{e}^{\frac{-\mathrm{j}\pi(2mM+1)}{(M+2K-L)}} \\ &\neq 1, \end{aligned} \tag{4.107}$$

for all m within the set $\{0, 1, \cdots, M+2K-L-1\}$, since $\frac{2mM+1}{M+2K-L}$ is not an even number.

for all m within the set $\{0, 1, \cdots, M+2K-L-1\}$, since $\frac{2mM+1}{M+2K-L}$ is not an even number.

Thus, a multicarrier system can be designed by setting $\overline{\mathbf{F}} = \mathbf{W}_M^H$ and $\overline{\mathbf{G}} = \overline{\mathbf{F}}^{-1}\mathbf{K} = \mathbf{W}_M\mathbf{K}$, yielding

$$\overline{\mathbf{G}} = \left[\sum_{r=0}^{3} \mathbf{D}_{\overline{\mathbf{p}}_r} \mathbf{W}_M \left[\mathbf{D}_M \quad \mathbf{0}_{M\times(2K-L)}\right] \mathbf{W}_{(M+2K-L)} \mathbf{D}^*_{\overline{\mathbf{q}}_r}\right] \mathbf{W}^H_{(M+2K-L)} \mathbf{D}^H_{(M+2K-L)}, \quad (4.109)$$

where the definitions of vectors $\overline{\mathbf{p}}_r$ and $\overline{\mathbf{q}}_r$ depend on whether the ZF or the MMSE solution is chosen. In any case, the resulting multicarrier structure is depicted in Figure 4.3.

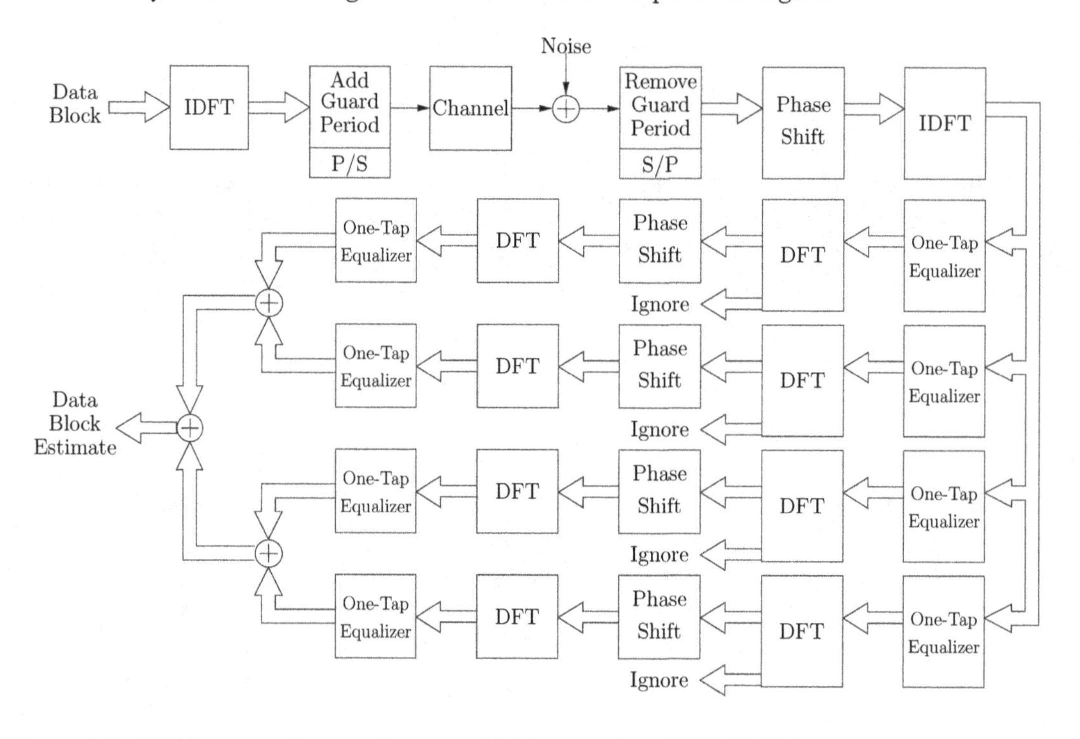

Figure 4.3: Multicarrier reduced-redundancy block transceiver (MC-RRBT).

Figure 4.3 illustrates that, after removing the guard period, there are $M+2K-L$ phase shifts to be performed in parallel, where the mth phase rotation is implemented through the multiplication by $e^{j\frac{\pi}{M(M+2K-L)}m}$. The first equalization step is performed after applying the IDFT to the data vector. The resulting data vector is simultaneously processed by four different branches at the receiver. The equalizers at this point consist of single-tap equalizers. The coefficients of the equalizers are the entries of vectors $\overline{\mathbf{q}}_r^*$, with $r \in \{0, 1, 2, 3\}$. After the DFT application, the last $2K-L$ elements are discarded and new phase shifts are performed on the remaining M elements, where in this case the mth rotation consists of a multiplication by $e^{-j\frac{\pi}{M^2}m}$. The final stage of the equalization consists of applying DFTs in parallel to the output signal of the rotators, followed by another single-tap equalization stage whose coefficients are the entries of vectors $\overline{\mathbf{p}}_r$, with $r \in \{0, 1, 2, 3\}$.

Also, a single-carrier system can be designed by setting $\overline{\mathbf{F}} = \mathbf{I}_M$ and $\overline{\mathbf{G}} = \overline{\mathbf{F}}^{-1}\mathbf{K} = \mathbf{K}$, yielding

$$\overline{\mathbf{G}} = \mathbf{W}_M^H \left[\sum_{r=0}^{3} \mathbf{D}_{\overline{\mathbf{p}}_r} \mathbf{W}_M \left[\mathbf{D}_M \quad \mathbf{0}_{M\times(2K-L)} \right] \mathbf{W}_{(M+2K-L)} \mathbf{D}_{\overline{\mathbf{q}}_r}^* \right] \mathbf{W}_{(M+2K-L)}^H \mathbf{D}_{(M+2K-L)}^H, \tag{4.110}$$

in which, once again, the definitions of vectors $\overline{\mathbf{p}}_r$ and $\overline{\mathbf{q}}_r$ depend on whether the ZF or the MMSE solution is chosen.

The superfast multicarrier and single-carrier transceivers of this chapter yield an additional degree of freedom in the ZP-ZJ-based transmissions, since the amount of redundancy can vary from the minimum value, $\lceil L/2 \rceil$, to the most commonly used value, L.[18] Nonetheless, one must deal with two distinct DFT sizes, M and $M + 2K - L$. When M is a power of 2, then $M + 2K - L$ is not necessarily a power of two. Thus, a radix-2 FFT algorithm could only be applied to implement those DFTs with size M. As for the DFTs with size $M + 2K - L$, one could implement the operations using a radix-2 FFT of size $2M$ (which is assumed to be larger than $M + 2K - L$), along with zero-padding of the related signals. Another possibility is to choose the amount of redundant elements in such a way that $M + 2K - L$ can be decomposed as a product of small prime numbers, leading to fast implementations as well. This is an open research topic.

4.5.1 COMPLEXITY COMPARISONS

This subsection describes some possible complexity comparisons between OFDM-based systems and superfast ZP-ZJ systems. Firstly, assume that an FFT algorithm requires $\frac{M}{2}\log_2 M - \frac{3M}{2} + 2$ complex-valued multiplications for processing size-M data blocks, as described in [17]. In addition, consider that $L = \frac{M}{4}$, as performed in [55]. In this case, it is possible to show that the number of complex-valued multiplications required by the superfast multicarrier reduced-redundancy system, by the overlap-and-add (OLA), and by fast proposals of zero-padded OFDM systems described in [55] is given by:

- ZP-OFDM-OLA: $M\log_2 M - 2M + 4$.
- ZP-OFDM-FAST: $\frac{5M}{4}\log_2 M - 5M + 20$.
- MC-RRBT: $\frac{15M}{2}\log_2 M - \frac{9M}{2} + 20 + 5(2K - L)$.

In the MC-RRBT, it is possible to implement part of the receiver side using parallel processing (see Figure 4.3). In this case, if we consider that the time that is required to perform a generic complex-valued multiplication is $T \in \mathbb{R}_+$ s, then the MC-RRBT requires $T(3M\log_2 M + 2(2K - L) + 8)$ s,

[18] It is worth highlighting that Figure 4.3 is a way to implement general ZP-OFDM systems ($K = L$) which allows for superfast equalization of received data blocks.

whereas the ZP-OFDM-OLA requires $T(M \log_2 M - 2M + 4)$ s, and the ZP-OFDM-FAST requires $T\left(\frac{5M}{4}\log_2 M - 5M + 20\right)$ s.

It is worth noting that, in the minimum-redundancy case where $K = L/2$ (assuming a channel with even order), the ZF solution can be further simplified. Indeed, as $2K - L = 0$, then $\mathbf{K}_{\text{ZF}} = \overline{\mathbf{H}}^{-1}$, implying that Equation (4.97) can be rewritten as $\mathbf{\mathcal{Q}} = [\mathbf{0}_{M\times 2}\ \ \mathbf{K}_{\text{ZF}}^H\hat{\mathbf{\mathcal{Q}}}]$. As the product $\mathbf{P}\mathbf{\mathcal{Q}}^H$ defines completely the displacement matrix $\nabla_{\mathbf{Z}_\xi,\mathbf{Z}_\eta}(\mathbf{K}_{\text{ZF}})$, then we can also discard the first two columns of $\mathbf{P}$, since they will be multiplied by zero anyway. Thus, for ZF minimum-redundancy systems, we can rewrite Equations (4.96) and (4.97) as follows:

$$\mathbf{P} = -\mathbf{K}_{\text{ZF}}\hat{\mathbf{P}}, \tag{4.111}$$

$$\mathbf{\mathcal{Q}} = \mathbf{K}_{\text{ZF}}^H\hat{\mathbf{\mathcal{Q}}}. \tag{4.112}$$

Such a result could be alternatively obtained by using Equation (4.50). An important consequence of this result is that all DFT matrices are $M \times M$ now and that the number of equalizer branches at the receiver end changes from four to two branches.

We assumed that the pair of generator matrices $(\mathbf{P}, \mathbf{\mathcal{Q}})$ is known. In fact, these matrices completely define the reduced-redundancy equalizers, since they are the only ones that contain information about the channel. These matrices, however, must be previously computed in the so-called *receiver-design stage*, which could be performed using up to $\mathcal{O}(M \log_2^2 M)$ operations, as described, for example, in [51]. Besides, there are many applications in which the receiver-design problem is not frequently solved. In wireline communications systems, the channel model is not updated so often. In this case, the main computational burden refers to the *equalization* itself.

Reducing the amount of operations in the receiver-design stage is still an interesting and rather important research direction in the context of superfast reduced-redundancy systems based on the ZP-ZJ model.

4.5.2 EXAMPLES

Up to this point, we showed efficient structures for reduced-redundancy systems allowing the equalization of received data blocks in a superfast manner. Nevertheless, we should note that ZP-ZJ systems do not necessarily outperform OFDM-based transceivers. In fact, the authors in [54] have proved mathematically that the MSE performance of reduced-redundancy transceivers degrades as the number of transmitted redundant elements reduces. That is, transmissions using larger amounts of redundancy leads to lower average MSE of symbols in ZP-ZJ systems.

If on one hand, we want to reduce the transmitted redundancy in order to save bandwidth; on the other hand, we need to use as much redundancy as possible in order to have a good MSE performance. The throughput is a good figure of merit to study the tradeoff between bandwidth usage and error performance. Thus, let us analyze the throughput performance of superfast ZP-ZJ systems through some simulation examples.

Example 4.4 (Minimum-Redundancy Transceivers) The figure of merit adopted here is the throughput, defined as

$$\text{Throughput} = br_c \frac{M}{M+K}(1 - \text{BLER}) f_s \quad \text{bps}, \tag{4.113}$$

in which b denotes the number of bits that is required to represent one constellation symbol, r_c denotes the code rate considering the protection of channel coding, K denotes the amount of redundancy, f_s denotes the sampling frequency, where symbol and channel models use the same sampling frequency, and BLER stands for block-error rate, assuming that a data block is discarded when at least one of its original bits is incorrectly decoded at the receiver end. In addition, the definition of the SNR used throughout the simulations is the ratio between the mean energy of the transmitted symbols at the input of the multipath channel and the power-spectral density of the additive noise at the receiver front-end. Besides, we also consider that both synchronization and channel estimation are perfectly performed at the receiver end.

In this example, we transmit 200 blocks, each one containing $M = 32$ BPSK data symbols (without taking redundancy into account), and compute the throughput by using a Monte-Carlo averaging process with 10,000 simulations. Consider these symbols are sampled at a frequency $f_s = 1.0$ GHz and that they are transmitted through a channel with a model operating at the same frequency as the symbols and with a long impulse response of order $L = 30$.[19] All the channel taps have the same variance, and the channel model is always normalized, that is, $E\left[\|\mathbf{h}\|_2^2\right] = 1$. Both the imaginary and real parts of the channel are independently drawn from a white and Gaussian random sequence (random Rayleigh channel). For each simulation a new channel is generated.

Furthermore, since the ZP-ZJ transceivers use zeros as redundant elements, the adopted OFDM and SC-FD systems in the simulations are the ZP-OFDM-OLA and ZP-SC-FD-OLA [55], where ZP and OLA stand for zero-padding and overlap-and-add, respectively, (see Subsections 2.4.2 and 2.4.4). Like the traditional cyclic-prefix-based systems, these ZP-based transceivers also induce a circulant channel matrix. We chose these transceivers as benchmarks since they are superfast transceivers that transmit L redundant zeros for each M data symbols. In summary, from now on we shall consider that OFDM means ZP-OFDM-OLA and SC-FD means ZP-SC-FD-OLA in all results here.

Figures 4.4 and 4.5 show the throughput curves for the OFDM, the SC-FD, the multi-carrier minimum-redundancy block transceiver (MC-MRBT), and the single-carrier minimum-redundancy block transceiver (SC-MRBT), using both ZF and MMSE designs. By observing these figures it is possible to verify that the minimum-redundancy superfast transceivers outperform the traditional OFDM and SC-FD transceivers in this particular setup, except for SNRs lower than

[19] We consider that $L = 30$ corresponds to a long impulse response since L is very close to M in this case.

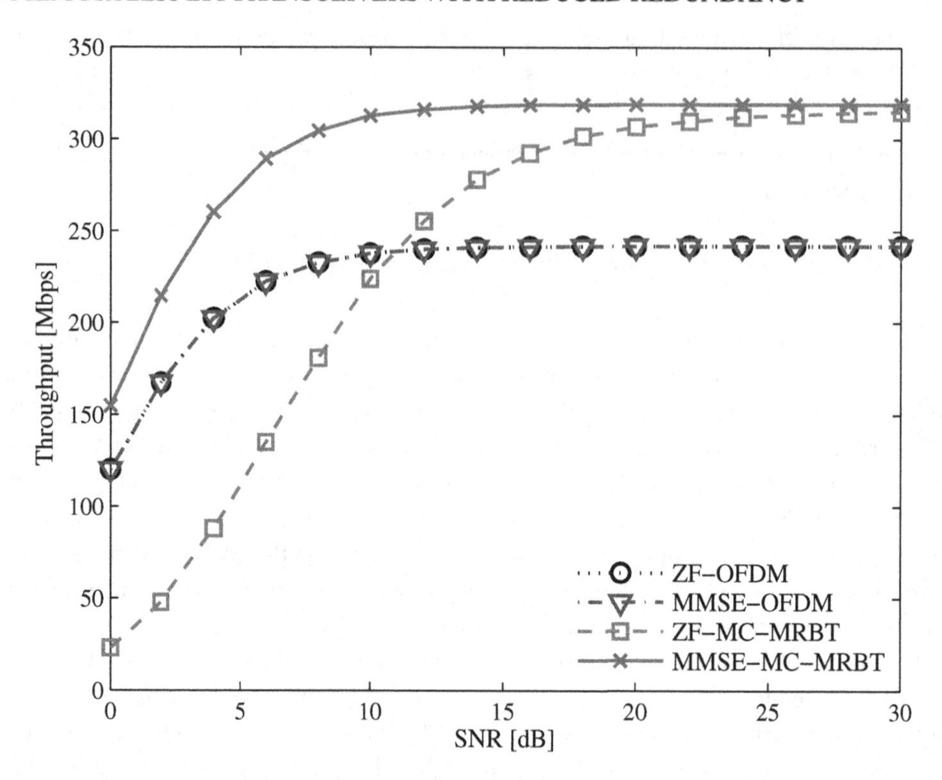

Figure 4.4: Throughput [Mbps] as a function of SNR [dB] for random Rayleigh channels, considering multicarrier transmissions.

12 dB in the ZF solutions. In this example, we use a convolutional code with constraint length 7, $r_c = 1/2$, and generators $\mathbf{g_0} = [\mathbf{133}]$ (octal) and $\mathbf{g_1} = [\mathbf{165}]$ (octal). This configuration is adapted from the LTE (long-term evolution) specifications [94]. In addition, for the BLER computation, we consider that a block (16 bits) is lost if, at least, one of its received bits is incorrect. We have employed a MATLAB implementation of a hard-decision Viterbi decoder. Note that such favorable result stems from the choices for M and L, representing delay constrained applications in very dispersive environment. These types of applications are suitable for the ZP-ZJ transceivers. In the cases where $M \gg L$, the traditional OFDM and SC-FD solutions are more adequate. □

Example 4.5 (Reduced-Redundancy Transceivers) In Example 4.4, we have shown that minimum-redundancy systems may significantly improve the throughput performance of multicarrier and single-carrier transmissions. Nevertheless, minimum-redundancy transceivers may also incur in high noise enhancements induced by the "inversion" of the Toeplitz effective channel matrix in the equalization process. In this example, we first chose a fourth-order channel model (see [83], pp. 306–307)

$$H_{\mathrm{A}}(z) \triangleq 0.1659 + 0.3045z^{-1} - 0.1159z^{-2} - 0.0733z^{-3} - 0.0015z^{-4} \tag{4.114}$$

for which the throughput performance of the minimum-redundancy systems is poor. For this channel (*Channel A*), we transmit 50,000 data blocks carrying $M = 16$ symbols of a 64-QAM constellation ($b = 6$ bits per symbol). In fact, each data block stems from 48 data bits that, after channel coding, yield 96 bits to be baseband modulated. The channel coding is the same as in Example 4.4 and we assume that the sampling frequency is $f_s = 100$ MHz.

Figure 4.6 depicts the obtained throughput results. We compare four different transceivers: the ZP-OFDM-OLA and the three possible multicarrier reduced-redundancy block transceivers (MC-RRBT). There are three possible MC-RRBT systems since the amount of redundant elements respects the inequality $\frac{L}{2} \leq K \leq L$ (i.e., $K \in \{2, 3, 4\}$). From Figure 4.6, one can observe that the minimum-redundancy multicarrier system (MC-RRBT for $K = 2$) that employs an MMSE equalizer is not able to produce a reliable estimate for the transmitted bits. However, if just one

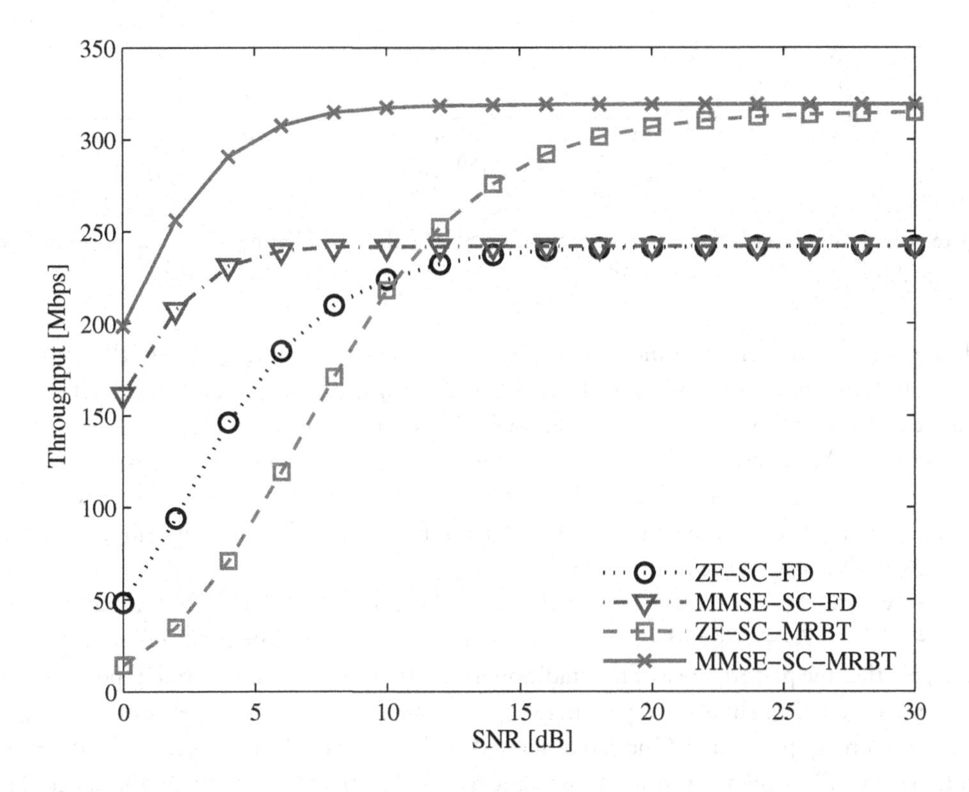

Figure 4.5: Throughput [Mbps] as a function of SNR [dB] for random Rayleigh channels, considering single-carrier transmissions.

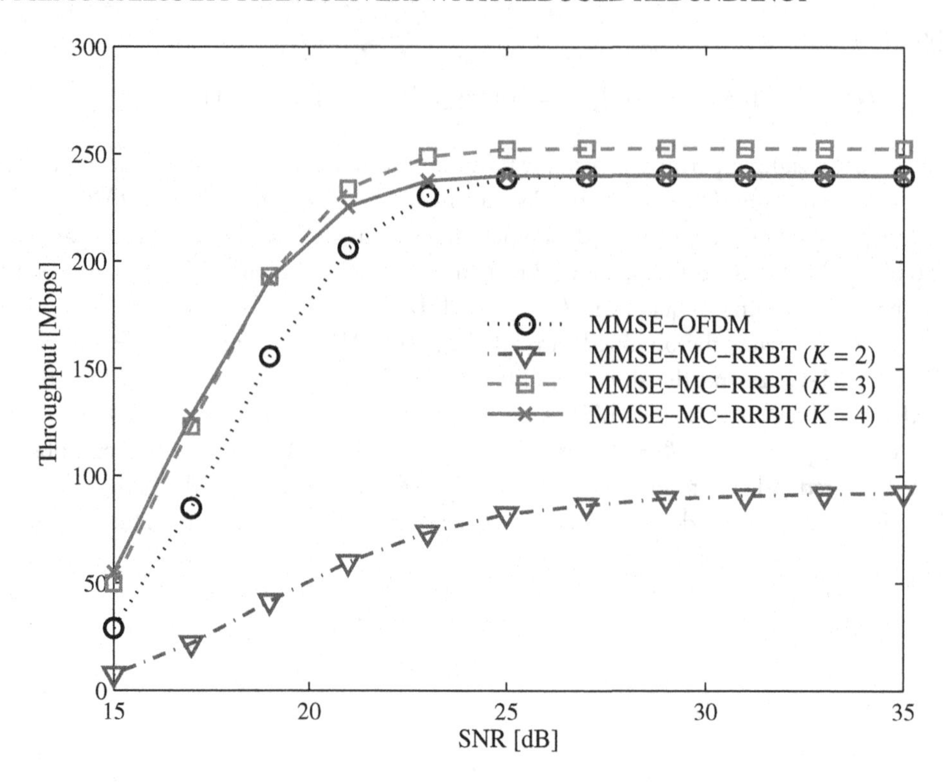

Figure 4.6: Throughput [Mbps] as a function of SNR [dB] for Channel A, considering multicarrier transmissions.

additional redundant element is included in the transmission, the resulting MC-RRBT system ($K = 3$) is enough to outperform the MMSE-OFDM. One should bear in mind that such throughput gains are attained without increasing substantially the computational complexity related to OFDM-based systems. Moreover, the MC-RRBT system with $K = 3$ also outperforms the MC-RRBT system with $K = 4$ in terms of throughput, especially for large SNR values, i.e., adding another redundant element in the transmission (MC-RRBT for $K = 4$) does not contribute to improving the throughput performance in this case.

Now, we will consider an FIR-channel model (*Channel B*) whose zeros are 0.999, −0.999, 0.7j, −0.7j, and −0.4j. This channel has zeros very close to the unit circle. We therefore expect that the performance of the traditional SC-FD system should be rather poor. Apart from the channel model, all simulation parameters are the same of the previous experiment. Figure 4.7 depicts the throughput results. One can observe that the SC-RRBT systems always outperform the traditional SC-FD system. Another important fact is that the throughput performance does not necessarily improves as the number of transmitted redundant elements is increased. For example, for low SNR values, it is better to use a reduced-redundancy system that transmits with a large number of

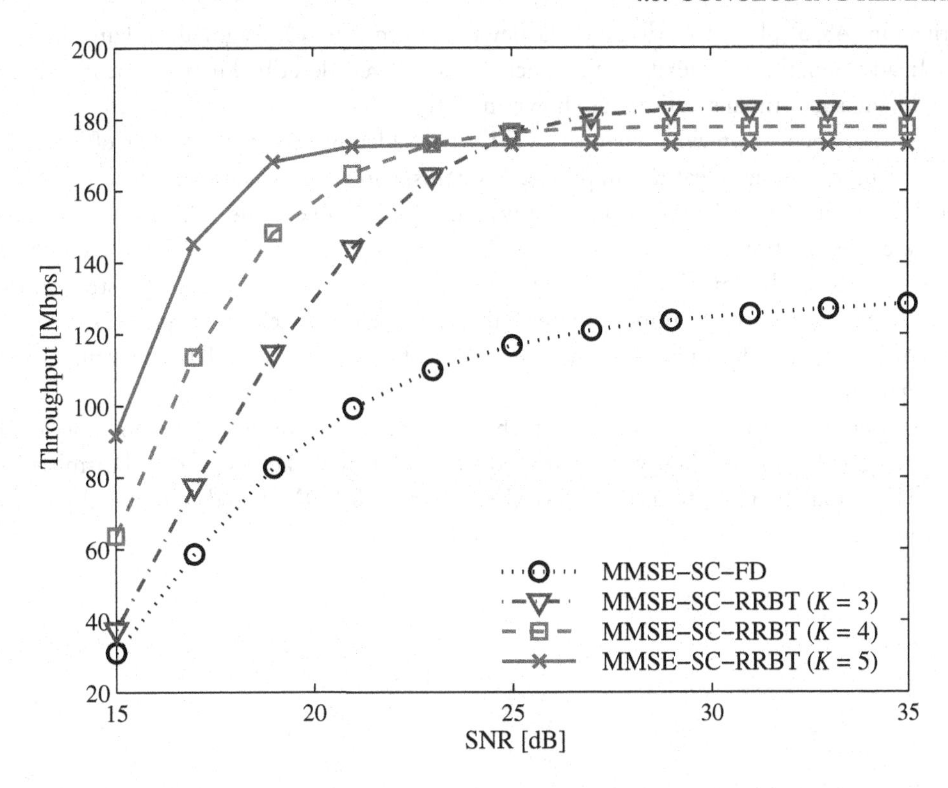

Figure 4.7: Throughput [Mbps] as a function of SNR [dB] for Channel B, considering single-carrier transmissions.

redundant elements ($K = 5$), whereas for large SNR values, it is better to use a reduced-redundancy system that transmits with a small number of redundant elements ($K = 3$). Once again, it is important to highlight that the superfast ZP-ZJ systems described in this chapter are just examples of how to transmit with a small number of redundant elements while using superfast transforms and single-tap equalizers. □

4.6 CONCLUDING REMARKS

This chapter described how to design memoryless LTI single-carrier and multicarrier transceivers with reduced redundancy for both ZF and MMSE optimal receivers. The block transceivers presented here are computationally efficient. We also introduced the mathematical tools which allow structured matrices to be represented through displacement operators. In particular, we emphasized the representation of Bezoutian matrices employing DFT, IDFT, and diagonal matrices. It is worth mentioning that similar decomposition is possible by using transforms with real entries such as those described in [48, 52] based on discrete Hartley transform (DHT), diagonal, and antidiagonal matrices. In addition, the complexity of the efficient transceivers described here can be further reduced by employing suboptimal solutions, as shown in [50].

As described in previous chapters, OFDM-based transceivers are rather efficient systems incorporating two desired features in practical systems, namely good/fair performance and computational simplicity. Nevertheless, there is always room for improvements. We showed in this chapter how to increase the transmission data-rates in block transceivers, while keeping the computational complexity close to the complexity of OFDM-based systems. We believe that the strategy followed in this chapter is even more important than the described transceivers themselves, since it illustrates how one can deduce new systems using a set of efficient tools related to structured matrix representations.

An important question still remains: what happens when we move from memoryless LTI to time-varying FIR systems? Is it possible to reduce even further the amount of transmitted redundancy? These and other questions related to what is beyond OFDM-based systems will be addressed in Chapter 5.

CHAPTER 5

FIR LTV Transceivers with Reduced Redundancy

5.1 INTRODUCTION

Nowadays, many practical communications systems employ the orthogonal frequency-division multiplexing (OFDM) as their core physical-layer modulation, as previously highlighted in Section 1.3. Such widespread adoption is due to many good properties that OFDM-based systems enjoy, as thoroughly explained in Chapter 3. However, one of the main drawbacks of OFDM is its related loss of spectral efficiency caused by the insertion of redundant elements in the transmission. This is an important issue considering the current trend of increasing the demand for data transmissions which shows no sign of settling. For example, the amount of wireless data services is more than doubling each year [37] leading to spectrum shortage as a sure event soon. It turns out that all efforts to maximize the spectrum usage are therefore highly justifiable at this point.

The memoryless linear time-invariant (LTI) transceivers with reduced redundancy described in Chapter 4 are a possible way of tackling the problem of increasing the spectral efficiency. Indeed, we show in Chapter 4 that, as compared to OFDM-based systems, reduced-redundancy LTI systems may decrease the amount of transmitted redundant elements in up to fifty percent, thus allowing a better use of the available spectrum for transmissions. However, this amount of reduction may still not be enough, especially in delay-constrained applications in very dispersive environments, in which the size of the transmitted data block cannot be too large and, in addition, the channel model is lengthy. Hence, we should ask ourselves if the memoryless LTI-based solutions could be further improved.

The transceivers described in Chapters 2, 3, and 4 are concrete examples of memoryless, linear, and time-invariant systems. Linearity plays a central role in the overall system design since it yields simpler transceivers, a very desirable feature in practice. A step ahead would be to investigate how to design simple time-varying transceivers with memory implemented through time-varying finite-impulse response (FIR) filters.[1] The memory and time-varying properties would introduce additional degrees of freedom in the design of systems that could help us circumvent some limitations inherent to memoryless LTI transceivers.

In fact, the time-varying FIR transceiver increases the transmission diversity, since more than one version of the transmitted message is received due to the memory of the system, and, in addition,

[1] In this case, we consider that the orders of the related filters are larger than or equal to 1. A memoryless system can be regarded as an FIR system whose order is equal to 0.

these versions are more likely to differ from each other due to the time-varying characteristic. If the receiver can take these facts into account, then it could trade off the amount of transmitted redundant elements with the amount of memory and degree of time-variance in the systems. Moreover, FIR linear time-varying (LTV) transceivers allow the design of high-selective transmitter filters and the interpretation of code-division multiple access (CDMA) schemes with long codes as a particular type of LTV transmultiplexer (TMUX), as will be explained later on.

It is worth recalling that intersymbol interference (ISI) is one of the most harmful effects inherent to broadband communications. That is why communication engineers often constrain their designs to eliminate ISI, especially when noise and other types of interference are negligible, leading to the so-called *zero-forcing* (ZF) transceivers. This chapter will describe some conditions that FIR LTV systems must satisfy in order to achieve ZF solutions. It turns out that some redundant elements must always be introduced in the ZF designs. On the other hand, ZF systems can be regarded as suboptimal solutions in the mean square error (MSE) sense. In this case, we can give up imposing the ZF constraint upon the transceiver, thus allowing the transmission without adding any redundant element, as long as *pure minimum MSE-based solutions* are employed. By pure minimum MSE (MMSE) we mean that there is no ZF constraint imposed upon the transceiver design.

Hence, the way to reduce even further the amount of redundancy inserted in the physical layer is either to design LTV transceivers and/or to allow the transmitter and receiver multiple-input multiple-output (MIMO) filters to have memory. In such cases, it is possible to derive ZF solutions *using only one redundant element.* Although time-varying transceivers with memory bring about a reduction in redundancy to allow for ZF solutions, their fast implementations are not known. In addition, the numerical accuracy of transceivers with reduced redundancy is not fully exploited in the open literature and for certain is a crucial issue to be addressed before any attempt to include these solutions in practical implementations or standards.

This chapter adopts a rather different approach as compared to Chapters 2, 3, and 4. The previous chapters focus on the derivation/description of either structures or performances of OFDM and beyond-OFDM systems, whereas the main focus of this chapter is on the limits of some parameters related to FIR LTV transceivers. We start by describing carefully how to use our previous knowledge about FIR LTI transceivers developed in Chapter 2 in order to model FIR LTV systems (Section 5.2). Then, we study the fundamental limits concerning some parameters inherent to FIR LTV transceivers that satisfy the ZF constraint, namely memory of receiver MIMO matrix and number of transmitted redundant elements (Section 5.3). After showing that when no redundancy is employed in the transmission ZF solutions cannot be achieved, then we present pure MMSE-based solutions (Section 5.4). Some examples are also given in this chapter (Section 5.5), including the interpretation of the ZF conditions within the framework of CDMA systems with long codes.

5.2 TIME-VARYING REDUCED-REDUNDANCY SYSTEMS WITH MEMORY

The time-varying property of FIR LTV systems hinders the direct application of the $\mathcal{Z}$-domain tools described in Chapter 2. Hence, in order to derive an appropriate model for the LTV transceivers, we will revisit the FIR LTI TMUX model introduced in Chapter 2 and derive its matrix representation in Subsection 5.2.1. Such matrix representation will be the key tool to describe the time-varying transceivers in Subsection 5.2.2.

5.2.1 FIR MIMO MATRICES OF LTI TRANSCEIVERS

Let us start with Figure 2.9 of Chapter 2, which depicts the $\mathcal{Z}$-domain description of block-based LTI transceivers. In this figure, $\mathbf{F}(z)$ represents an FIR MIMO transmitter matrix whose entries are polynomials in the complex variable z^{-1}, as given in Equation (2.42). If we assume that, among all entries of $\mathbf{F}(z)$, the highest polynomial order is $T-1$, with T being a given natural number,[2] then we can decompose $\mathbf{F}(z)$ as

$$\mathbf{F}(z) = \sum_{\tau \in \mathcal{T}} \mathbf{F}_\tau z^{-\tau}, \tag{5.1}$$

where $\mathcal{T} \triangleq \{0, \cdots, T-1\}$ is a set of integer indexes and $\mathbf{F}_\tau$ is an $N \times M$ matrix with complex-valued entries, rather than polynomial entries. The natural numbers M and N are, respectively, the number of symbols within a data block and the actual number of transmitted elements per block, which includes both data symbols and redundant elements. The entries of $\mathbf{F}_\tau$, for all $\tau \in \mathcal{T}$, can be determined directly from $\mathbf{F}(z)$ using Equation (5.1). However, these coefficients are not important in our discussions here.

A transmitter matrix with memory ($T > 1$) enables one to design high-selective transmitter filters, which may be very useful to avoid spectrum leakage (out-of-band transmissions), to reduce intercarrier interference (ICI) that occurs specially when carrier-frequency offset (CFO) is significant, and to compensate for the channel distortions when channel-state information (CSI) is available at the transmitter side. Besides, having memory at the transmitter increases the transmitter diversity, which could enhance the performance of the related transceiver.

Similarly to what we did with $\mathbf{F}(z)$, by assuming that the order of the FIR MIMO receiver matrix $\mathbf{G}(z)$ in Figure 2.9 is $Q-1$, where Q is a natural number, we can decompose $\mathbf{G}(z)$ as

$$\mathbf{G}(z) = \sum_{q \in \mathcal{Q}} \mathbf{G}_q z^{-q}, \tag{5.2}$$

where $\mathbf{G}_q$ is an $M \times N$ complex-valued matrix, for each integer index q within the set $\mathcal{Q} \triangleq \{0, \cdots, Q-1\}$. Once again, the entries of matrix $\mathbf{G}_q$ are not important in our discussions here, but

[2]From now on, we will define the *order of a matrix* as the highest polynomial order among all of its entries. For instance, the order of $\mathbf{F}(z)$ is $T-1$.

these entries are defined by $\mathbf{G}(z)$ which in turn is defined in Equation (2.43). It is worth mentioning that, if the transmitted redundant elements are not able to completely eliminate the interblock interference (IBI), then matrix $\mathbf{G}(z)$ must have memory, i.e., it is mandatory to have $Q > 1$ in some applications, as will be clearer soon.

Once again we will perform the same type of decomposition, but now for the FIR MIMO channel matrix $\mathbf{H}(z)$ in Figure 2.9. Thus, we are interested in representing $\mathbf{H}(z)$ as

$$\mathbf{H}(z) = \sum_{b \in \mathcal{B}} \mathbf{H}_b z^{-b}, \tag{5.3}$$

where $\mathbf{H}_b$ is an $N \times N$ complex-valued Toeplitz matrix, for each integer index b within the set $\mathcal{B} \triangleq \{0, 1, \cdots, B\}$, where B stands for the order of the channel matrix $\mathbf{H}(z)$. The entries of $\mathbf{H}_b$ can be determined by the entries of $\mathbf{H}(z)$, which in turn is a pseudo-circulant matrix defined by Equations (2.29) and (2.30) of Chapter 2. Following these steps, it is possible to show that the entries of $\mathbf{H}_b$ are related to the channel impulse response $h(l)$ of order L by means of the following equality:

$$\left[\mathbf{H}_b\right]_{ml} \triangleq h(bN + m - l), \tag{5.4}$$

for any indexes m and l within the set $\mathcal{N} \triangleq \{0, 1, \cdots, N-1\}$. In addition, since we assumed that the order of the causal channel is L, then $h(bN + m - l) = 0$ whenever $bN + m - l$ is larger than L or smaller than 0.

Example 5.1 (Determining $\mathbf{H}_b$ from $\mathbf{H}(z)$) Suppose the order L of the FIR channel impulse response $h(l)$ is such that $N < L < 2N$, with N being the length of the transmitted block. Thus, we are assuming that the channel memory L is longer than N. By using Equations (2.29) and (2.30) of Chapter 2, we will decompose $\mathbf{H}(z)$ according to Equation (5.3), as follows:

$$\mathbf{H}(z) = \underbrace{\begin{bmatrix} h(0) & 0 & 0 & \cdots & \cdots & \cdots & 0 \\ h(1) & h(0) & 0 & \cdots & \cdots & \cdots & 0 \\ h(2) & h(1) & h(0) & 0 & \cdots & \cdots & 0 \\ \vdots & \vdots & \vdots & \ddots & \ddots & \vdots & \vdots \\ h(N-3) & \vdots & \vdots & \ddots & \ddots & \ddots & \vdots \\ h(N-2) & h(N-3) & \vdots & \cdots & \cdots & h(0) & 0 \\ h(N-1) & h(N-2) & h(N-3) & \cdots & \cdots & \cdots & h(0) \end{bmatrix}}_{\triangleq \mathbf{H}_0 \in \mathbb{C}^{N \times N}}$$

$$
+z^{-1}\underbrace{\begin{bmatrix}
h(N) & h(N-1) & \cdots & h(4) & h(3) & h(2) & h(1) \\
h(N+1) & h(N) & \ddots & \ddots & h(4) & h(3) & h(2) \\
\vdots & h(N+1) & \ddots & \ddots & \ddots & \ddots & h(3) \\
\vdots & \vdots & \ddots & \ddots & \ddots & \ddots & h(4) \\
\vdots & \vdots & \ddots & \ddots & \ddots & \ddots & \vdots \\
\vdots & \vdots & \ddots & \ddots & \ddots & \ddots & h(N-1) \\
\cdots & \cdots & \cdots & \cdots & \cdots & h(N+1) & h(N)
\end{bmatrix}}_{\triangleq \mathbf{H}_1 \in \mathbb{C}^{N\times N}}
$$

$$
+z^{-2}\underbrace{\begin{bmatrix}
0 & \cdots & 0 & h(L) & \cdots & h(N+1) \\
0 & 0 & \cdots & 0 & \ddots & \vdots \\
\vdots & \vdots & \vdots & \vdots & \vdots & h(L) \\
0 & 0 & 0 & \cdots & 0 & 0 \\
0 & 0 & 0 & \cdots & 0 & 0 \\
\vdots & \vdots & \vdots & \vdots & \vdots & \vdots \\
0 & 0 & 0 & 0 & \cdots & 0
\end{bmatrix}}_{\triangleq \mathbf{H}_2 \in \mathbb{C}^{N\times N}}
$$

$$
\begin{aligned}
&= \mathbf{H}_0 + z^{-1}\mathbf{H}_1 + z^{-2}\mathbf{H}_2 \\
&= \sum_{b\in\mathcal{B}} \mathbf{H}_b z^{-b},
\end{aligned} \tag{5.5}
$$

with $B = 2$. In addition, note that the Toeplitz matrices $\mathbf{H}_b$ are such that $\mathbf{H}_0$ is a lower triangular matrix, whereas $\mathbf{H}_B = \mathbf{H}_2$ is an upper triangular matrix. □

Equation (5.4) together with Example 5.1 suggest that B, the order of $\mathbf{H}(z)$, can be related to the order of the channel impulse response L and the length of the transmitted block N. Now, let us analyze such relation.

When $L < N$, Equation (5.3) coincides with Equation (2.38) of Chapter 2, with $\mathbf{H}_0 = \mathbf{H}_{\text{ISI}}$ and $\mathbf{H}_1 = \mathbf{H}_{\text{IBI}}$, thus we have $B = 1$. When $L = N$, the decomposition in Equation (5.3) still requires only two matrices, i.e., we still have $B = 1$, but in this case $\mathbf{H}_1 = \mathbf{H}_{\text{IBI}}$ includes a non-zero diagonal entry $h(L) = h(N)$. Physically, this means that the entire transmitted block suffers interference from the previous transmitted block. As L becomes larger than N, other previously transmitted blocks start interfering with the current transmitted block and, therefore, additional matrices $\mathbf{H}_b$ appear in the decomposition of $\mathbf{H}(z)$. Example 5.1 illustrates the case when $N < L < 2N$, which leads to $B = 2$. In addition, when $2N < L \leq 3N$ we have $B = 3$. Hence, we can define

B as

$$B \triangleq \left\lceil \frac{L}{N} \right\rceil. \tag{5.6}$$

The definition of B in expression (5.6) implies that $BN \geq L$. This observation will be particularly useful in Section 5.3.

As previously observed, when the length of the transmitted data block N is smaller than the channel order L, the resulting interblock interference (IBI) is caused by more than one block that have been previously transmitted. This case ($B > 1$) is representative of delay-constrained applications whose related channel-impulse responses are very long, thus hindering the transmission of data blocks with length $N \geq L$ ($B = 1$). Another observation is that, since the channel order is L and we are considering causal impulse responses, then $h(L)$ must be necessarily different from zero. In addition, we will assume that the channel model has a non-zero first term in its impulse response, i.e., we assume $h(0) \neq 0$, which means that the channel transfer function has no zeros at the origin.

With those previous definitions, we can now derive a matrix description for the LTI transceiver of Figure 2.9. To do that, let us first assume that the $\mathcal{Z}$-transform of a generic time-domain vector signal $\mathbf{x}(n)$ is denoted as

$$\mathbf{x}(z) \triangleq \mathcal{Z}\left\{\mathbf{x}(n)\right\}, \tag{5.7}$$

where $\mathbf{x}(n)$ has size $N_x \in \mathbb{N}$ and can be expressed as

$$\mathbf{x}(n) \triangleq [\, x(nN_x - N_x + 1) \;\; x(nN_x - N_x + 2) \;\; \cdots \;\; x(nN_x)\,]^T. \tag{5.8}$$

Thus, in the time domain, considering a noiseless channel, i.e., $\mathbf{v}(n) = \mathbf{0}_{N \times 1}$ for all time index $n \in \mathbb{Z}$, we have the following expressions for the transmitted, received, and estimated vectors, respectively:

$$\begin{aligned}
\mathbf{u}(n) &\triangleq \mathcal{Z}^{-1}\left\{\mathbf{F}(z)\mathbf{s}(z)\right\} \\
&= \sum_{\tau \in \mathcal{T}} \mathbf{F}_\tau \mathcal{Z}^{-1}\left\{z^{-\tau}\mathbf{s}(z)\right\} \\
&= \sum_{\tau \in \mathcal{T}} \mathbf{F}_\tau \mathbf{s}(n - \tau), && (5.9) \\
\mathbf{y}(n) &\triangleq \mathcal{Z}^{-1}\left\{\mathbf{H}(z)\mathbf{u}(z)\right\} \\
&= \sum_{b \in \mathcal{B}} \mathbf{H}_b \mathcal{Z}^{-1}\left\{z^{-b}\mathbf{u}(z)\right\} \\
&= \sum_{b \in \mathcal{B}} \mathbf{H}_b \mathbf{u}(n - b), && (5.10) \\
\hat{\mathbf{s}}(n) &\triangleq \mathcal{Z}^{-1}\left\{\mathbf{G}(z)\mathbf{y}(z)\right\} \\
&= \sum_{q \in \mathcal{Q}} \mathbf{G}_q \mathcal{Z}^{-1}\left\{z^{-q}\mathbf{y}(z)\right\} \\
&= \sum_{q \in \mathcal{Q}} \mathbf{G}_q \mathbf{y}(n - q). && (5.11)
\end{aligned}$$

We can write the input-output relationship of FIR LTI transceivers in a much more compact form by conveniently stacking the variables which appear in Equations (5.9), (5.10), and (5.11). Indeed, note that Equation (5.11) can be rewritten as follows:

$$\begin{aligned}\hat{\mathbf{s}}(n) &= \underbrace{\begin{bmatrix}\mathbf{G}_{Q-1} & \cdots & \mathbf{G}_0\end{bmatrix}}_{\triangleq \mathcal{G} \in \mathbb{C}^{M \times QN}} \underbrace{\begin{bmatrix}\mathbf{y}\left(n-(Q-1)\right) \\ \vdots \\ \mathbf{y}(n)\end{bmatrix}}_{\triangleq \overline{\mathbf{y}}(n) \in \mathbb{C}^{QN \times 1}} \\ &= \mathcal{G}\overline{\mathbf{y}}(n). \end{aligned} \tag{5.12}$$

Equation (5.12) depends on vectors $\mathbf{y}\left(n-Q+1\right), \cdots, \mathbf{y}(n)$, which can be determined using Equation (5.10) as follows:

$$\begin{aligned}\overline{\mathbf{y}}(n) &= \underbrace{\begin{bmatrix}\mathbf{H}_B & \cdots & \mathbf{H}_0 & \mathbf{0} & \cdots & \mathbf{0} \\ \mathbf{0} & \mathbf{H}_B & \cdots & \mathbf{H}_0 & \cdots & \mathbf{0} \\ \vdots & \vdots & \ddots & & \ddots & \vdots \\ \mathbf{0} & \mathbf{0} & \cdots & \mathbf{H}_B & \cdots & \mathbf{H}_0\end{bmatrix}}_{\triangleq \mathcal{H} \in \mathbb{C}^{QN \times (B+Q)N}} \underbrace{\begin{bmatrix}\mathbf{u}\left(n-(B+Q-1)\right) \\ \vdots \\ \mathbf{u}(n-1) \\ \mathbf{u}(n)\end{bmatrix}}_{\triangleq \overline{\mathbf{u}}(n) \in \mathbb{C}^{(B+Q)N \times 1}} \\ &= \mathcal{H}\overline{\mathbf{u}}(n). \end{aligned} \tag{5.13}$$

Equation (5.13), on its turn, depends on $\mathbf{u}\left(n-B-Q+1\right), \cdots, \mathbf{u}(n)$, which can be computed using Equation (5.9) as follows:

$$\begin{aligned}\overline{\mathbf{u}}(n) &= \underbrace{\begin{bmatrix}\mathbf{F}_{T-1} & \cdots & \mathbf{F}_0 & \mathbf{0} & \cdots & \mathbf{0} \\ \mathbf{0} & \mathbf{F}_{T-1} & \cdots & \mathbf{F}_0 & \cdots & \mathbf{0} \\ \vdots & \vdots & \ddots & & \ddots & \vdots \\ \mathbf{0} & \mathbf{0} & \cdots & \mathbf{F}_{T-1} & \cdots & \mathbf{F}_0\end{bmatrix}}_{\triangleq \mathcal{F} \in \mathbb{C}^{(B+Q)N \times [B+Q+(T-1)]M}} \underbrace{\begin{bmatrix}\mathbf{s}\left(n-(B+Q+T-2)\right) \\ \vdots \\ \mathbf{s}(n-1) \\ \mathbf{s}(n)\end{bmatrix}}_{\triangleq \overline{\mathbf{s}}(n) \in \mathbb{C}^{[B+Q+(T-1)]M \times 1}} \\ &= \mathcal{F}\overline{\mathbf{s}}(n). \end{aligned} \tag{5.14}$$

Using Equations (5.12), (5.13), and (5.14), we have the following overall input-output description of the block-based LTI transceiver of Figure 2.9:

$$\hat{\mathbf{s}}(n) = \mathcal{G}\mathcal{H}\mathcal{F}\overline{\mathbf{s}}(n), \tag{5.15}$$

which is the time-domain version of equation

$$\hat{\mathbf{s}}(z) = \mathbf{G}(z)\mathbf{H}(z)\mathbf{F}(z)\mathbf{s}(z) \tag{5.16}$$

in the $\mathcal{Z}$-domain. This transceiver is time-invariant since the *global transmitter and receiver matrices*, $\mathcal{F}$ and $\mathcal{G}$, respectively, are constant matrices that do not depend on the time index n. The above matrix description is rather useful for the time-varying case, since we cannot apply the $\mathcal{Z}$-domain description of the TMUX of Chapter 2 for this case.

5.2.2 FIR MIMO MATRICES OF LTV TRANSCEIVERS

As previously mentioned, time-variance brings about extra degrees of freedom that play the role of introducing additional diversity to the system. In order to generalize the previous LTI model to the LTV case, we could consider that each transmitter matrix $\mathbf{F}_\tau$, with $\tau \in \mathcal{T}$, is actually a time-varying matrix $\mathbf{F}_\tau(n)$, which depends on the time index n. This implies that matrices $\mathbf{F}_0(n), \cdots, \mathbf{F}_{T-1}(n)$ are employed to generate the channel input $\mathbf{u}(n)$, whereas matrices $\mathbf{F}_0(n-1), \cdots, \mathbf{F}_{T-1}(n-1)$ are employed to generate the channel input $\mathbf{u}(n-1)$, and so forth. Similarly, we could consider that the receiver matrix $\mathbf{G}_q$, with $q \in \mathcal{Q}$, is in fact a matrix $\mathbf{G}_q(n)$, which varies with time as well.

In summary, by using this simple generalization we can rewrite Equations (5.9), (5.10), and (5.11) as follows:

$$\mathbf{u}(n) = \sum_{\tau \in \mathcal{T}} \mathbf{F}_\tau(n)\mathbf{s}(n-\tau), \tag{5.17}$$

$$\mathbf{y}(n) = \sum_{b \in \mathcal{B}} \mathbf{H}_b\mathbf{u}(n-b), \tag{5.18}$$

$$\hat{\mathbf{s}}(n) = \sum_{q \in \mathcal{Q}} \mathbf{G}_q(n)\mathbf{y}(n-q). \tag{5.19}$$

With these new equations it is straightforward to define generalized time-varying transmitter and receiver matrices, $\boldsymbol{\mathcal{F}}(n)$ and $\boldsymbol{\mathcal{G}}(n)$, as follows:

$$\boldsymbol{\mathcal{G}}(n) \triangleq \begin{bmatrix}\mathbf{G}_{Q-1}(n) & \cdots & \mathbf{G}_0(n)\end{bmatrix}, \tag{5.20}$$

$$\boldsymbol{\mathcal{F}}(n) \triangleq \begin{bmatrix} \mathbf{F}_{T-1}(n-Q-B+1) & \cdots & \mathbf{F}_0(n-Q-B+1) & \mathbf{0} & \cdots & \mathbf{0} \\ \vdots & \ddots & \ddots & & \ddots & \vdots \\ \mathbf{0} & \mathbf{0} & \cdots & \mathbf{F}_{T-1}(n) & \cdots & \mathbf{F}_0(n) \end{bmatrix}. \tag{5.21}$$

Hence, we have the following estimate at the receiver end:

$$\hat{\mathbf{s}}(n) = \boldsymbol{\mathcal{G}}(n)\boldsymbol{\mathcal{H}}\boldsymbol{\mathcal{F}}(n)\bar{\mathbf{s}}(n), \tag{5.22}$$

and due to the time dependency of the transceiver matrices,[3] there is no $\mathcal{Z}$-domain counterpart for such equation.

The reader should observe that, due to the definition of $\bar{\mathbf{s}}(n)$ in Equation (5.14), we have

$$\mathbf{s}(n) = \begin{bmatrix}\mathbf{0}_{M\times(B+Q+T-2)M} & \mathbf{I}_M\end{bmatrix}\bar{\mathbf{s}}(n), \tag{5.23}$$

implying that the ZF solution is achieved, i.e., $\hat{\mathbf{s}}(n) = \mathbf{s}(n)$, whenever

$$\boldsymbol{\mathcal{G}}(n)\boldsymbol{\mathcal{H}}\boldsymbol{\mathcal{F}}(n) = \begin{bmatrix}\mathbf{0}_{M\times(B+Q+T-2)M} & \mathbf{I}_M\end{bmatrix}, \tag{5.24}$$

[3] For simplicity, we are considering that the channel does not vary with time. However, since we are interested in bounds rather than transceiver designs, we could decompose time-varying channels into two parts, one being time-varying and the other time invariant, and then incorporate the time-varying part of the channel in the transmitter matrix, as shown in [70]. Therefore, some conditions/bounds derived in this chapter are also applicable to time-varying channels.

or, in other words, whenever we have

$$\underbrace{\mathcal{F}^H(n)}_{(Q+B+T-1)M\times(Q+B)N} \times \underbrace{\mathcal{H}^H}_{(Q+B)N\times QN} \times \underbrace{\mathcal{G}^H(n)}_{QN\times M} = \begin{bmatrix} \mathbf{0}_{(B+Q+T-2)M\times M} \\ \mathbf{I}_M \end{bmatrix}. \tag{5.25}$$

The specific constraints that the transceiver matrices $\mathcal{F}(n)$ and $\mathcal{G}(n)$ must satisfy in order to achieve the ZF solution expressed in (5.25) will be described in the following section.

5.3 CONDITIONS FOR ACHIEVING ZF SOLUTIONS

The aim of this section is to examine the conditions that the global transceiver matrices $\mathcal{F}(n)$ and $\mathcal{G}(n)$ must satisfy in order for Equation (5.25) to hold. There are a variety of parameters that can affect such ZF constraint. Therefore, we shall fix some of them by first assuming that the order of the FIR MIMO transmitter matrix is already known, being denoted by $T-1$. In addition, we shall assume that the order L of the scalar channel $h(l)$ is also known. The number of symbols to be transmitted in a block was previously fixed at M. With those assumptions, our ultimate target is to find some mathematical bounds concerning the order of the FIR MIMO receiver matrix (Subsection 5.3.2), the number of redundant elements K that might possibly be inserted in the transmission, and, as a consequence, the length of the transmitted data block $N = M + K$ (Subsections 5.3.3 and 5.3.4), and the degree of time-variance of the transceiver (Subsection 5.3.5).

Before proceeding we will first rewrite the ZF constraint expressed in (5.25) in a different manner, as described in Subsection 5.3.1.

5.3.1 THE ZF CONSTRAINT

Let us start by interpreting Equation (5.25) as follows: for any $(B+Q+T-1)M \times 1$ canonical vector $\mathbf{e}_m$,[4] with m within the set $\{(B+Q+T-1)M-M, (B+Q+T-1)M-M+1, \cdots, (B+Q+T-1)M-1\}$, one has

$$\mathbf{e}_m \in \mathcal{R}\left\{\mathcal{F}^H(n)\mathcal{H}^H\mathcal{G}^H(n)\right\} \subset \mathcal{R}\left\{\mathcal{F}^H(n)\mathcal{H}^H\right\} \subset \mathcal{R}\left\{\mathcal{F}^H(n)\right\} \subset \mathbb{C}^{(B+Q+T-1)M}, \tag{5.26}$$

in which $\mathcal{R}\{\cdot\}$ denotes the *column* or *range space* of the argument.[5] This means that, for all such vectors $\mathbf{e}_m$, Equation (5.25) holds if there always exists a $QN \times 1$ vector $\boldsymbol{\beta}_m$ such that

$$\mathbf{e}_m = \mathcal{F}^H(n)\mathcal{H}^H\boldsymbol{\beta}_m, \tag{5.27}$$

with $m \in \{(B+Q+T-1)M-M, (B+Q+T-1)M-M+1, \cdots, (B+Q+T-1)M-1\}$.

By observing the previous equation, one can see that the solution to such linear system depends on how matrix $\mathcal{H}^H$ transforms the vector $\boldsymbol{\beta}_m$ into a new vector within the domain of $\mathcal{F}^H(n)$. An

[4]Here, $\mathbf{e}_m$ is a $(B+Q+T-1)M \times 1$ vector whose mth entry is equal to 1, whereas the remaining elements are equal to 0.

[5]The range space of a given matrix is the vector subspace spanned by linear combinations of the columns of the related matrix. It is the image of the linear mapping associated with such a matrix.

important fact concerning the range space of matrix $\boldsymbol{\mathcal{H}}^H$ is that, for all $(Q+B)N \times 1$ vector $\boldsymbol{\alpha}$ within the range space of $\boldsymbol{\mathcal{H}}^H$, one has

$$\boldsymbol{\alpha} = \begin{bmatrix} \mathbf{0}_{(BN-L)\times 1} \\ \overline{\boldsymbol{\alpha}} \end{bmatrix}, \tag{5.28}$$

where $\overline{\boldsymbol{\alpha}}$ is within $\mathcal{R}\left\{\overline{\boldsymbol{\mathcal{H}}}^H\right\} \subset \mathbb{C}^{QN+L}$, in which $\overline{\boldsymbol{\mathcal{H}}}^H$ is comprised of the last $QN+L$ rows of $\boldsymbol{\mathcal{H}}^H$. That is, we can construct matrix $\overline{\boldsymbol{\mathcal{H}}}^H$ by simply discarding the first $BN-L \geq 0$ rows of $\boldsymbol{\mathcal{H}}^H$ and keeping the remaining rows unchanged. Here, we consider that, if $BN-L=0$, then Equation (5.28) should be read as $\boldsymbol{\alpha} = \overline{\boldsymbol{\alpha}}$.

It remains to answer why Equation (5.28) holds. If $BN = L$, then there is nothing to be proved. Thus, assume that $BN > L$. By observing Equation (5.13), one can see that the only possibility of having non-zero entries in the first $BN-L$ rows of matrix $\boldsymbol{\mathcal{H}}^H$ occurs when there are some non-zero entries in the first $BN-L$ rows of matrix $\mathbf{H}_B^H$. Based on Equation (5.4), the first $BN-L$ rows of matrix $\mathbf{H}_B^H$ contains the elements $h^*(BN+m-l)$, where $m \in \mathcal{N}$ denotes the column index and $l \in \{0, \cdots, BN-L-1\}$ denotes the row index. However, $BN+m-l$ is always larger than L for such indexes m and l (indeed, the smallest argument is achieved when $m=0$ and $l = BN-L-1$, thus yielding $BN+m-l = L+1 > L$), which implies that $h^*(BN+m-l)=0$. As a consequence, all linear combinations of the columns of $\boldsymbol{\mathcal{H}}^H$ have their first $BN-L$ entries equal to zero, whereas the remaining entries are corresponding linear combinations of the columns of $\overline{\boldsymbol{\mathcal{H}}}^H$. This means that Equation (5.28) is always true. Example 5.2 illustrates the above reasoning.

Example 5.2 (Range Space of Matrix $\boldsymbol{\mathcal{H}}^H$) Let us start with the same pseudo-circulant channel matrix $\mathbf{H}(z)$ and its related matrices $\mathbf{H}_0$, $\mathbf{H}_1$, and $\mathbf{H}_2$ of Example 5.1, where $N < L < 2N$, thus implying that $B=2$. In this example, we will study the range space of matrix $\boldsymbol{\mathcal{H}}^H$. From expression (5.13), we can write

$$\begin{aligned}\boldsymbol{\mathcal{H}}^H &\triangleq \begin{bmatrix} \mathbf{H}_2 & \mathbf{H}_1 & \mathbf{H}_0 & \mathbf{0} & \cdots & \mathbf{0} \\ \mathbf{0} & \mathbf{H}_2 & \mathbf{H}_1 & \mathbf{H}_0 & \cdots & \mathbf{0} \\ \vdots & \vdots & \ddots & & \ddots & \vdots \\ \mathbf{0} & \mathbf{0} & \cdots & \mathbf{H}_2 & \mathbf{H}_1 & \mathbf{H}_0 \end{bmatrix}^H \\ &= \begin{bmatrix} \mathbf{H}_2^H & \mathbf{0} & \cdots & \mathbf{0} \\ \mathbf{H}_1^H & \mathbf{H}_2^H & \cdots & \mathbf{0} \\ \mathbf{H}_0^H & \mathbf{H}_1^H & \ddots & \vdots \\ \mathbf{0} & \mathbf{H}_0^H & \ddots & \mathbf{H}_2^H \\ \vdots & \vdots & \ddots & \mathbf{H}_1^H \\ \mathbf{0} & \mathbf{0} & \cdots & \mathbf{H}_0^H \end{bmatrix},\end{aligned} \tag{5.29}$$

with $\mathbf{H}_2^H \in \mathbb{C}^{N\times N}$ being a lower triangular Toeplitz matrix given by

$$\mathbf{H}_2^H \triangleq \begin{bmatrix} 0 & \cdots & 0 & 0 & \cdots & 0 \\ \vdots & \vdots & \vdots & \vdots & \vdots & 0 \\ 0 & 0 & \cdots & 0 & \ddots & \vdots \\ h^*(L) & 0 & 0 & \cdots & 0 & 0 \\ h^*(L-1) & h^*(L) & 0 & \cdots & 0 & 0 \\ \vdots & \vdots & \ddots & \vdots & \vdots & \vdots \\ h^*(N+1) & h^*(N+2) & \cdots & h^*(L) & \cdots & 0 \end{bmatrix}. \tag{5.30}$$

Therefore, every entry corresponding to the first $BN - L = 2N - L$ rows of matrix $\mathbf{H}_2^H$ is equal to 0. Consequently, the first $BN - L$ rows of $\boldsymbol{\mathcal{H}}^H$ are also comprised of 0 elements only. Hence, any vector formed by linear combination of the columns of $\boldsymbol{\mathcal{H}}^H$ will have zeros in its first $BN - L$ entries, as shown in Equation (5.28). □

Taking Equation (5.28) into account and by defining $\boldsymbol{\alpha}_m \triangleq \boldsymbol{\mathcal{H}}^H \boldsymbol{\beta}_m$, we have that the linear system of Equation (5.27) can be rewritten as

$$\begin{aligned} \mathbf{e}_m &= \boldsymbol{\mathcal{F}}^H(n)\boldsymbol{\alpha}_m \\ &= \boldsymbol{\mathcal{F}}^H(n) \underbrace{\begin{bmatrix} \mathbf{0}_{(BN-L)\times 1} \\ \overline{\boldsymbol{\alpha}}_m \end{bmatrix}}_{\triangleq \boldsymbol{\alpha}_m} \\ &= \overline{\boldsymbol{\mathcal{F}}}^H(n)\overline{\boldsymbol{\alpha}}_m, \end{aligned} \tag{5.31}$$

in which $\overline{\boldsymbol{\mathcal{F}}}^H(n)$ is a $(Q+B+T-1)M \times (QN+L)$ matrix containing the last $QN + L$ columns of matrix $\boldsymbol{\mathcal{F}}^H(n)$ (i.e., its first $BN - L$ columns are discarded).

Therefore, we must have at least one solution $\overline{\boldsymbol{\alpha}}_m$ within the column space of matrix $\overline{\boldsymbol{\mathcal{H}}}^H$, i.e., $\overline{\boldsymbol{\alpha}}_m \in \mathcal{R}\left\{\overline{\boldsymbol{\mathcal{H}}}^H\right\} \subset \mathbb{C}^{(QN+L)\times 1}$. From linear algebra, we know that if $\overline{\boldsymbol{\alpha}}_m$ is a vector within the column space of matrix $\overline{\boldsymbol{\mathcal{H}}}^H$, then $\overline{\boldsymbol{\alpha}}_m$ must be orthogonal to all vectors belonging to the *kernel* (null space) of matrix $\overline{\boldsymbol{\mathcal{H}}}$, since the subspace $\ker\left\{\overline{\boldsymbol{\mathcal{H}}}\right\}$ is the *orthogonal complement* of the subspace $\mathcal{R}\left\{\overline{\boldsymbol{\mathcal{H}}}^H\right\}$, in which $\ker\{\cdot\}$ denotes the kernel of the argument.[6] Hence, we can restate the ZF constraint as satisfying the linear system in expression (5.31), while also respecting the constraint

$$\mathbf{z}^H\overline{\boldsymbol{\alpha}}_m = 0, \tag{5.32}$$

for all vector $\mathbf{z}$ within the subspace $\ker\left\{\overline{\boldsymbol{\mathcal{H}}}\right\}$, and for each m within the set $\{(B+Q+T-1)M - M, (B+Q+T-1)M - M + 1, \cdots, (B+Q+T-1)M - 1\}$.

[6]The kernel (or null space) of a given matrix is the vector subspace spanned by vectors that have zero image. An important result in linear algebra is that the dimension of the null space plus the dimension of the range space of a matrix is equal to its number of columns.

Now, we can use this interpretation of the ZF constraint in order to derive a lower bound for Q, the order of the MIMO receiver matrix described in Equation (5.19).

5.3.2 LOWER BOUND ON THE RECEIVER LENGTH

Assume that the $(Q+B+T-1)M \times (QN+L)$ matrix $\overline{\boldsymbol{\mathcal{F}}}^H(n)$ has been previously designed in such a way that it has rank $(Q+B+T-1)M-(BN-L)$. This might sound a bit artificial, but we shall illustrate it later on, in a concrete example in Subsection 5.3.4. By now it is sufficient to know the motivation for such a choice. The main idea is that we must have a rank large enough to allow us to solve M linear systems such as the one described in Equation (5.31). It turns out that the number of degrees of freedom induced by a rank of $(Q+B+T-1)M-(BN-L)$ is more than enough to solve the related linear systems. Such an assumption requires that

$$QN+L \geq (Q+B+T-1)M-(BN-L). \tag{5.33}$$

From linear algebra we know that the dimension of the null space of $\overline{\boldsymbol{\mathcal{F}}}^H(n)$ is

$$QN+L-[(Q+B+T-1)M-(BN-L)] = (Q+B)K-(T-1)M, \tag{5.34}$$

in which the integer number

$$K \triangleq N-M \tag{5.35}$$

denotes the amount of redundancy introduced in the transmission.

The rank constraint upon matrix $\overline{\boldsymbol{\mathcal{F}}}^H(n)$ implies that there is a $(QN+L)\times[(Q+B)K-(T-1)M]$ matrix $\boldsymbol{\Theta}(n)$, whose rows span the subspace $\ker\left\{\overline{\boldsymbol{\mathcal{F}}}^H(n)\right\}$. Thus, for all given vector $\mathbf{e}_m$, there always exists at least one particular solution $\hat{\boldsymbol{\gamma}}_m$ such that $\mathbf{e}_m = \overline{\boldsymbol{\mathcal{F}}}^H(n)\hat{\boldsymbol{\gamma}}_m$. The general solution to the linear system in Equation (5.31) can therefore be written as

$$\overline{\boldsymbol{\alpha}}_m = \boldsymbol{\Theta}(n)\hat{\boldsymbol{\alpha}} + \hat{\boldsymbol{\gamma}}_m, \tag{5.36}$$

where $\hat{\boldsymbol{\alpha}} \in \mathbb{C}^{[(Q+B)K-(T-1)M]\times 1}$ is a vector which parameterizes the general solution and introduces the additional degrees of freedom that allow $\overline{\boldsymbol{\alpha}}_m$ to satisfy the requirement of being orthogonal to all vectors within the null space of $\overline{\boldsymbol{\mathcal{H}}}$ (see Equation (5.31)). Note that

$$\begin{aligned}
\overline{\boldsymbol{\mathcal{F}}}^H(n)\overline{\boldsymbol{\alpha}}_m &= \underbrace{\overline{\boldsymbol{\mathcal{F}}}^H(n)\boldsymbol{\Theta}(n)}_{=\mathbf{0}}\hat{\boldsymbol{\alpha}} + \underbrace{\overline{\boldsymbol{\mathcal{F}}}^H(n)\hat{\boldsymbol{\gamma}}_m}_{=\mathbf{e}_m} \\
&= \mathbf{0}\times\hat{\boldsymbol{\alpha}} + \mathbf{e}_m \\
&= \mathbf{e}_m.
\end{aligned} \tag{5.37}$$

Now, consider the restriction of having $\overline{\boldsymbol{\alpha}}_m \in \mathcal{R}\left\{\overline{\boldsymbol{\mathcal{H}}}^H\right\}$, which is equivalent to find $\overline{\boldsymbol{\alpha}}_m$ such that $\mathbf{z}^H\overline{\boldsymbol{\alpha}}_m = 0$ for all vector $\mathbf{z}$ within the subspace $\ker\left\{\overline{\boldsymbol{\mathcal{H}}}\right\}$. But, instead of considering any vector

$\mathbf{z} \in \ker\left\{\overline{\mathcal{H}}\right\}$, we can verify if $\overline{\boldsymbol{\alpha}}_m$ is orthogonal to all vectors from a basis of $\ker\left\{\overline{\mathcal{H}}\right\}$. Indeed, if the dimension of the kernel of matrix $\overline{\mathcal{H}}$ is $D \in \mathbb{N}$, then there are D linear independent vectors $\mathbf{z}_0, \mathbf{z}_1, \cdots, \mathbf{z}_{D-1}$ which span $\ker\left\{\overline{\mathcal{H}}\right\}$ and which respect the identity $\overline{\mathcal{H}}\mathbf{z}_d = \mathbf{0}_{QN\times 1}$, for all d within the set $\mathcal{D} \triangleq \{0, 1, \cdots, D-1\}$. Since any vector $\mathbf{z} \in \overline{\mathcal{H}}$ can be written as

$$\mathbf{z} = \zeta_0\mathbf{z}_0 + \cdots + \zeta_{D-1}\mathbf{z}_{D-1}, \tag{5.38}$$

then

$$\mathbf{z}^H\overline{\boldsymbol{\alpha}}_m = \zeta_0\mathbf{z}_0^H\overline{\boldsymbol{\alpha}}_m + \cdots + \zeta_{D-1}\mathbf{z}_{D-1}^H\overline{\boldsymbol{\alpha}}_m. \tag{5.39}$$

Hence, if $\mathbf{z}_d^H\overline{\boldsymbol{\alpha}}_m = 0$ for all d within the set $\mathcal{D}$, then $\mathbf{z}^H\overline{\boldsymbol{\alpha}}_m = 0$. Thus, our task is to find a basis for $\ker\left\{\overline{\mathcal{H}}\right\}$. It turns out that such a basis is associated with the zeros of the scalar channel transfer function, as formulated below.

Given the FIR channel model, whose transfer function is

$$H(z) \triangleq h(0) + h(1)z^{-1} + \cdots + h(L)z^{-L}, \tag{5.40}$$

with $h(0) \neq 0 \neq h(L)$, consider that the zeros of $H(z)$ are all distinct, i.e., there are L distinct scalars $z_0, \cdots, z_{L-1} \in \mathbb{C}$ such that $H(z_0) = \cdots = H(z_{L-1}) = 0$. Since $h(0) \neq 0 \neq h(L)$ and due to the Toeplitz structure of matrix $\mathcal{H}$ described in Equation (5.13), it turns out that this channel matrix is full-row rank, i.e., rank of $\mathcal{H}$ is QN. As $\overline{\mathcal{H}}$ is generated from $\mathcal{H}$ by discarding the first $BN - L$ zero columns, then the rank of $\overline{\mathcal{H}}$ is QN as well. This means that the dimension of the null space of $\overline{\mathcal{H}}$ is

$$\begin{aligned} D &= QN + L - QN \\ &= L. \end{aligned} \tag{5.41}$$

Moreover, the set composed by vectors of the form

$$\mathbf{z}_l^T \triangleq \begin{bmatrix} 1 & z_l & z_l^2 & \cdots & z_l^{QN+L-1} \end{bmatrix}^T, \tag{5.42}$$

for each $l \in \{0, \cdots, L-1\}$, is a linearly independent set of vectors within the kernel of $\overline{\mathcal{H}}$, since

$$\overline{\mathcal{H}}\mathbf{z}_l = \mathbf{0}_{QN\times 1}, \tag{5.43}$$

for all $l \in \{0, \cdots, L-1\}$. Indeed, the ith row of $\overline{\mathcal{H}}$, with $i \in \{0, 1, \cdots, QN-1\}$, is given by

$$\overline{\mathbf{h}}_i^T \triangleq \begin{bmatrix} 0 & \cdots & 0 & h(L) & \cdots & h(0) & 0 \cdots 0 \end{bmatrix} \in \mathbb{C}^{1\times(QN+L)}, \tag{5.44}$$

where there are i zero entries at the beginning of this vector and $QN - i - 1$ zero entries at its end. Therefore, we have

$$\begin{aligned}
\overline{\mathbf{h}}_i^T \mathbf{z}_l &= h(L)z_l^i + h(L-1)z_l^{i+1} + \cdots + h(0)z_l^{L+i} \\
&= z_l^{L+i}\left[h(L)z_l^{-L} + h(L-1)z_l^{-(L-1)} + \cdots + h(0)\right] \\
&= z_l^{L+i} \underbrace{H(z_l)}_{=0} \\
&= 0,
\end{aligned} \tag{5.45}$$

implying that $\overline{\mathcal{H}}\mathbf{z}_l = \mathbf{0}_{QN\times 1}$, for any $l \in \{0, \cdots, L-1\}$.

Thus, the constraint described in Equation (5.32) can be rewritten as

$$\mathbf{Z}^H \overline{\boldsymbol{\alpha}}_m = \mathbf{0}_{L\times 1}, \tag{5.46}$$

where the $(QN + L) \times L$ full column-rank matrix $\mathbf{Z}$ is defined as

$$\mathbf{Z} \triangleq \begin{bmatrix} \mathbf{z}_0 & \mathbf{z}_1 & \cdots & \mathbf{z}_{L-1} \end{bmatrix}. \tag{5.47}$$

Hence, we can substitute the general solution described in Equation (5.36) into the constraint in Equation (5.46), thus giving

$$\mathbf{Z}^H \boldsymbol{\Theta}(n)\hat{\boldsymbol{\alpha}} + \mathbf{Z}^H \hat{\boldsymbol{\gamma}}_m = \mathbf{0}_{L\times 1}, \tag{5.48}$$

yielding

$$\left[\mathbf{Z}^H \boldsymbol{\Theta}(n)\right]\hat{\boldsymbol{\alpha}} = -\mathbf{Z}^H \hat{\boldsymbol{\gamma}}_m. \tag{5.49}$$

A condition[7] for achieving the solution is having

$$\hat{\boldsymbol{\alpha}} = -\left[\mathbf{Z}^H \boldsymbol{\Theta}(n)\right]^H \left[\left[\mathbf{Z}^H \boldsymbol{\Theta}(n)\right]\left[\mathbf{Z}^H \boldsymbol{\Theta}(n)\right]^H\right]^{-1} \mathbf{Z}^H \hat{\boldsymbol{\gamma}}_m, \tag{5.50}$$

which is achievable only when matrix $\left[\mathbf{Z}^H \boldsymbol{\Theta}(n)\right]\left[\mathbf{Z}^H \boldsymbol{\Theta}(n)\right]^H$ is non-singular. This fact has two consequences:

1. we must have $L \leq (Q + B)K - (T - 1)M$; and
2. the rank of the $L \times [(Q + B)K - (T - 1)M]$ matrix $\mathbf{Z}^H \boldsymbol{\Theta}(n)$ is in fact L.

[7] Such a constraint may not be a necessary condition, which means that we may derive some results more restrictive than the necessary ones.

The first consequence of the aforementioned rank requirement is that the number of columns of matrix $\left[\mathbf{Z}^H\boldsymbol{\Theta}(n)\right]\left[\mathbf{Z}^H\boldsymbol{\Theta}(n)\right]^H$ cannot be smaller than its number of rows. Therefore, we must have

$$(Q+B)K \geq (T-1)M+L, \tag{5.51}$$

yielding

$$Q \geq \left\lceil \frac{(T-1)M+L}{K} \right\rceil - B, \tag{5.52}$$

which is a lower bound on Q that must be used in the design of time-varying receiver matrix (remember that, by hypothesis, $Q \geq 1$).

5.3.3 LOWER BOUND ON THE AMOUNT OF REDUNDANCY

In order to achieve the sufficient solution of Equation (5.50), one must always introduce redundancy, i.e., one must have $K = N - M \geq 1$. In fact, we have already used this result in order to derive expression (5.52) from expression (5.51). Indeed, by examining inequality (5.51), we know that $(T-1)M + L \geq L \geq 1$, by hypothesis. Besides, as $Q + B$ is always larger than or equal to 2, then $K = N - M$ cannot be smaller than or equal to zero 0. Hence, we must necessarily have

$$K \geq 1. \tag{5.53}$$

Actually, having $N < M$ is not meaningful anyway since in this case the transmitter discards some data which are supposed to be transmitted, constituting a loss of information. However, when a design criteria other than the ZF solution is employed, one may consider using $N = M$ (i.e., $K = 0$) without conceptual problems (see Subsection 5.4 for further details on this topic).

Example 5.3 (Equalizer-Length Bound) This example is just a numerical illustration of the bound described in expression (5.52). Let us first assume that the transmitter is memoryless ($T = 1$), that $M \geq L/2$, and that $K = L/2$ (even order). Using these values we can deduce that $B = \lceil \frac{L}{M+L/2} \rceil = 1$ and, by using inequality (5.52), we get $Q \geq 1$, i.e., the receiver can be memoryless as well. See Chapter 4 for the designs of such memoryless transceivers.

In the case where we have an FIR transmitter matrix with $T = 3$, and assuming $K = 2$ and $L = 2N$, then $B = \lceil \frac{2N}{N} \rceil = 2$ and $Q \geq N + M - 2$, i.e., the time-varying receiver might have long memory. □

The reader should observe that the equalizer-length bound in expression (5.52) depends on some known variables, namely L, M, and T, and on the amount of redundancy K which is to be defined. Given L, M, and T, one might argue if it is always possible to satisfy inequality (5.52) and the constraint rank $\left\{\mathbf{Z}^H\boldsymbol{\Theta}(n)\right\} = L$ (see discussions after expression (5.50)), regardless of the amount of transmitted redundant elements K. The next subsection shows a particular example of time-varying transceiver with memory that is able to achieve the lower bound on K in expression (5.53) and discusses further design constraints to achieve such lower bound.

5.3.4 ACHIEVING THE LOWER BOUND OF REDUNDANCY

Let us address this topic by first considering for the time being that the transmitter matrices are memoryless ($T = 1$), since having more columns in matrix $\boldsymbol{\mathcal{F}}(n)$ could only add more diversity (more redundancy) to the system, so that the memoryless transmitter case is somehow the worst case scenario for finding a lower bound on K which is actually required to achieve the ZF solution. Indeed, it is reasonable to expect that the minimum achievable bound[8] for K would also be sufficient when we have memory at the transmitter end.

The remaining of this section will focus on proving that the lower bound on the number of transmitted redundant elements in expression (5.53) can be reached and we encourage the reader to keep this target in mind while examining the forthcoming lengthy mathematical computations. Another important observation is that we are not interested in the performance of the described transceiver, rather we focus only on the parameter bounds.

In the memoryless transmitter case, we consider that the time-varying transmitter matrix in its Hermitian transpose version is given by

$$\boldsymbol{\mathcal{F}}^H(n) \triangleq \begin{bmatrix} \mathbf{F}_0^H(n-Q-B+1) & \mathbf{0} & \cdots & \mathbf{0} \\ \mathbf{0} & \ddots & \cdots & \mathbf{0} \\ \vdots & \vdots & \mathbf{F}_0^H(n-1) & \vdots \\ \mathbf{0} & \mathbf{0} & \cdots & \mathbf{F}_0^H(n) \end{bmatrix} \in \mathbb{C}^{(Q+B)M\times(Q+B)N}, \quad (5.54)$$

where we assume that $\boldsymbol{\mathcal{F}}^H(n)$ has full row rank, which means that the rank of $\boldsymbol{\mathcal{F}}^H(n)$ is $(Q+B)M$. In addition, assume that, for each time index n, we have

$$\mathbf{F}_0(n) = \begin{bmatrix} \mathbf{I}_M \\ [\mathbf{0}_{K\times(BN-L)} \quad \boldsymbol{\Phi}(n)] \end{bmatrix} \mathbf{F}(n) \in \mathbb{C}^{N\times M}, \quad (5.55)$$

where $\mathbf{F}(n)$ is an $M \times M$ matrix with full rank, whereas $\boldsymbol{\Phi}(n)$ is a given $K \times \overline{M}$ matrix, with $\overline{M} \triangleq M - (BN - L) \geq 0$.[9] Such structure for the transmitter matrices means that the redundant elements are included at the end of each block to be transmitted and that these redundant elements are linear combinations of the last $\overline{M}$ non-redundant elements.

Now, matrix $\overline{\boldsymbol{\mathcal{F}}}^H(n)$ can be generated by discarding the first $BN - L$ columns of $\boldsymbol{\mathcal{F}}^H(n)$. Based on expressions (5.54) and (5.55), this means that we are removing $BN - L$ linearly independent columns of $\boldsymbol{\mathcal{F}}^H(n)$, thus reducing the rank from $(Q+B)M$ to $(Q+B)M - (BN-L) = (Q+B-1)M + M - (BN-L) = (Q+B-1)M + \overline{M}$. This coincides with the proposed dimension for $\overline{\boldsymbol{\mathcal{F}}}^H(n)$ described in the first paragraph of Subsection 5.3.2 (for $T = 1$). Therefore, we

[8]The term "achievable" was employed here since we do not know a priori if, for any channel model and for any choices of M and T, we could use $K = 1$ and still satisfy the constraint: rank $\left\{\mathbf{Z}^H\boldsymbol{\Theta}(n)\right\} = L$.

[9]This assumption ($\overline{M} \geq 0$) is not a very restrictive constraint, since if $B = 1$, then $BN - L \leq BN - K = N - K = M$ (assuming $L \geq K$), whereas if $B > 1$, then N can be chosen so that BN is close to L, implying that it is easy to satisfy $BN - L \leq M$.

have the following $(Q+B)M \times (QN+L)$ matrix:

$$\overline{\mathcal{F}}^H(n) \triangleq \begin{bmatrix} \mathbf{F}^H(n-Q-B+1) & \mathbf{0} & \cdots & \mathbf{0} \\ \mathbf{0} & \ddots & \cdots & \mathbf{0} \\ \vdots & \vdots & \mathbf{F}^H(n-1) & \vdots \\ \mathbf{0} & \mathbf{0} & \cdots & \mathbf{F}^H(n) \end{bmatrix}$$
$$\times \begin{bmatrix} \begin{bmatrix} \mathbf{0}_{(BN-L)\times\overline{M}} & \mathbf{0}_{(BN-L)\times K} \\ \mathbf{I}_{\overline{M}} & \mathbf{\Phi}^H(n-Q-B+1) \end{bmatrix} & \mathbf{0} \\ \mathbf{0} & \operatorname{diag}\left\{ \begin{bmatrix} \mathbf{I}_M & \begin{bmatrix} \mathbf{0}_{(BN-L)\times K} \\ \mathbf{\Phi}^H(n-i) \end{bmatrix} \end{bmatrix} \right\}_{i=Q+B-2}^{0} \end{bmatrix}. \tag{5.56}$$

From linear algebra, we know that the kernel of matrix $\overline{\mathcal{F}}^H(n)$ has dimension $QN+L-[(Q+B)M-(BN-L)] = (Q+B)K$. Let $\mathbf{\Theta}(n)$ be a $(QN+L)\times(Q+B)K$ block-diagonal matrix defined as

$$\mathbf{\Theta}(n) \triangleq \begin{bmatrix} \overline{\mathbf{\Psi}}(n-Q-B+1) & \mathbf{0} & \cdots & \mathbf{0} \\ \vdots & \ddots & \vdots & \vdots \\ \mathbf{0} & \cdots & \mathbf{\Psi}(n-1) & \mathbf{0} \\ \mathbf{0} & \mathbf{0} & \cdots & \mathbf{\Psi}(n) \end{bmatrix}, \tag{5.57}$$

where

$$\mathbf{\Psi}(n-i) \triangleq \begin{bmatrix} \begin{bmatrix} \mathbf{0}_{(BN-L)\times K} \\ -\mathbf{\Phi}^H(n-i) \end{bmatrix} \\ \mathbf{I}_K \end{bmatrix} \in \mathbb{C}^{N\times K}, \tag{5.58}$$

for all $i \in \{0, \cdots, Q+B-2\}$, whereas

$$\overline{\mathbf{\Psi}}(n-Q-B+1) \triangleq \begin{bmatrix} -\mathbf{\Phi}^H(n-Q-B+1) \\ \mathbf{I}_K \end{bmatrix} \in \mathbb{C}^{(\overline{M}+K)\times K}. \tag{5.59}$$

With this definition, one always has $\overline{\mathcal{F}}^H(n)\mathbf{\Theta}(n) = \mathbf{0}_{(Q+B)M\times(Q+B)K}$, which allows us to write the matrix equation

$$\overline{\boldsymbol{\alpha}}_m = \mathbf{\Theta}(n)\hat{\boldsymbol{\alpha}} + \hat{\boldsymbol{\gamma}}_m, \tag{5.60}$$

which is a concrete example of Equation (5.36), where $\hat{\boldsymbol{\gamma}}_m$ is a particular solution of the linear system $\overline{\mathcal{F}}^H(n)\overline{\boldsymbol{\alpha}}_m = \mathbf{e}_m$. Observe that, in order to determine $\overline{\boldsymbol{\alpha}}_m$, it is only necessary to determine $\hat{\boldsymbol{\alpha}}$, since $\hat{\boldsymbol{\gamma}}_m$ and $\mathbf{\Theta}(n)$ are assumed to be already known, since they depend exclusively on matrix $\overline{\mathcal{F}}(n)$. Let us also simplify the forthcoming notation by omitting the index m of the related variables. Thus, $\overline{\boldsymbol{\alpha}} = \mathbf{\Theta}(n)\hat{\boldsymbol{\alpha}} + \hat{\boldsymbol{\gamma}}$ must be a solution to the linear system $\mathbf{e} = \overline{\mathcal{F}}^H(n)\overline{\boldsymbol{\alpha}}$.

As the solution to such linear system must also satisfy $\mathbf{Z}^H\overline{\boldsymbol{\alpha}} = \mathbf{0}_{L\times 1}$, where the $L \times (QN + L)$ matrix $\mathbf{Z}^H$ is defined in Equation (5.47), then we have $\left[\mathbf{Z}^H\boldsymbol{\Theta}(n)\right]\hat{\boldsymbol{\alpha}} = -\mathbf{Z}^H\hat{\boldsymbol{\gamma}}$. The former equation can be rewritten in a much more convenient way by first defining vectors $\tilde{\mathbf{z}}_l$, for each $l \in \{0, \cdots, L-1\}$, as

$$\tilde{\mathbf{z}}_l^T \triangleq \begin{bmatrix} 1 & z_l & z_l^2 & \cdots & z_l^{N-1} \end{bmatrix}^T, \tag{5.61}$$

and then by rewriting the $L \times (Q + B)K$ matrix $\mathbf{Z}^H\boldsymbol{\Theta}(n)$ as follows:

$$\mathbf{Z}^H\boldsymbol{\Theta}(n) = \begin{bmatrix} \tilde{\mathbf{z}}_0^H\boldsymbol{\Psi}(n-Q-B+1) & \cdots & (z_0^*)^{(Q+B-2)N}\tilde{\mathbf{z}}_0^H\boldsymbol{\Psi}(n-1) & (z_0^*)^{(Q+B-1)N}\tilde{\mathbf{z}}_0^H\boldsymbol{\Psi}(n) \\ \tilde{\mathbf{z}}_1^H\boldsymbol{\Psi}(n-Q-B+1) & \cdots & (z_1^*)^{(Q+B-2)N}\tilde{\mathbf{z}}_1^H\boldsymbol{\Psi}(n-1) & (z_1^*)^{(Q+B-1)N}\tilde{\mathbf{z}}_1^H\boldsymbol{\Psi}(n) \\ \vdots & \vdots & \cdots & \vdots \\ \tilde{\mathbf{z}}_{L-1}^H\boldsymbol{\Psi}(n-Q-B+1) & \cdots & (z_{L-1}^*)^{(Q+B-2)N}\tilde{\mathbf{z}}_{L-1}^H\boldsymbol{\Psi}(n-1) & (z_{L-1}^*)^{(Q+B-1)N}\tilde{\mathbf{z}}_{L-1}^H\boldsymbol{\Psi}(n) \end{bmatrix} \tag{5.62}$$

where $\boldsymbol{\Psi}(n-Q-B+1)$ is an $N \times K$ matrix defined as

$$\boldsymbol{\Psi}(n-Q-B+1) \triangleq \begin{bmatrix} \mathbf{0}_{(BN-L)\times K} \\ \hline \overline{\boldsymbol{\Psi}}(n-Q-B+1) \end{bmatrix}. \tag{5.63}$$

Now, a possible condition for finding a solution for $\hat{\boldsymbol{\alpha}}$, as in Equation (5.50), is having $(Q + B)K \geq L$, or

$$K \geq \left\lceil \frac{L}{Q+B} \right\rceil. \tag{5.64}$$

It is worth pointing out that, in the memoryless receiver case ($Q = 1$) with $N \geq L$ (i.e., $B = 1$), condition (5.64) becomes $K \geq \left\lceil \frac{L}{2} \right\rceil$, which is the same ZF condition previously deduced in Chapter 4.

By observing the structure of matrix $\mathbf{Z}^H\boldsymbol{\Theta}(n)$ one can verify its strong dependency on the quantities z_l^N, for $l \in \{0, \cdots, L-1\}$, where z_l is a zero of the scalar channel transfer function $H(z)$. The number of degrees of freedom associated with that matrix is reduced when there exist distinct indexes l_1, l_2 such that $z_{l_1}^N = z_{l_2}^N$. In such cases, the chances of matrix $\mathbf{Z}^H\boldsymbol{\Theta}(n)$ to have full row rank is somehow decreased. In order to deal with such worst case setup, let us define the *congruous zeros* of $H(z)$ with respect to N as the distinct complex numbers z_{l_1} and z_{l_2} which satisfy $z_{l_1}^N = z_{l_2}^N \triangleq c \in \mathbb{C}$.

Thus, assume that the first $\mu \in \{1, \cdots, L\}$ zeros of $H(z)$ are congruous. Then, the first μ rows of $\mathbf{Z}^H\boldsymbol{\Theta}(n)$ can be rewritten as

$$\widehat{\mathbf{Z}^H\boldsymbol{\Theta}}(n) = \begin{bmatrix} \tilde{\mathbf{z}}_0^H\boldsymbol{\Psi}(n-Q-B+1) & \cdots & (c^*)^{Q+B-2}\tilde{\mathbf{z}}_0^H\boldsymbol{\Psi}(n-1) & (c^*)^{Q+B-1}\tilde{\mathbf{z}}_0^H\boldsymbol{\Psi}(n) \\ \vdots & \vdots & \cdots & \vdots \\ \tilde{\mathbf{z}}_{\mu-1}^H\boldsymbol{\Psi}(n-Q-B+1) & \cdots & (c^*)^{Q+B-2}\tilde{\mathbf{z}}_{\mu-1}^H\boldsymbol{\Psi}(n-1) & (c^*)^{Q+B-1}\tilde{\mathbf{z}}_{\mu-1}^H\boldsymbol{\Psi}(n) \end{bmatrix}$$
$$= \underbrace{\begin{bmatrix} \tilde{\mathbf{z}}_0^H \\ \vdots \\ \tilde{\mathbf{z}}_{\mu-1}^H \end{bmatrix}}_{\mu\times N} \underbrace{\begin{bmatrix} \boldsymbol{\Psi}(n-Q-B+1) & \cdots & \boldsymbol{\Psi}(n-1) & \boldsymbol{\Psi}(n) \end{bmatrix}}_{N\times(Q+B)K}$$
$$\times \underbrace{\begin{bmatrix} \mathbf{I}_K & & & \\ & c^*\mathbf{I}_K & & \\ & & \ddots & \\ & & & (c^*)^{Q+B-1}\mathbf{I}_K \end{bmatrix}}_{(Q+B)K\times(Q+B)K}. \tag{5.65}$$

Consider that the global transmitter matrix was designed in such a way that there exist $\mathbb{N} \ni \delta \leq Q+B$ matrices, let us say $\boldsymbol{\Psi}(n-\delta+1), \cdots, \boldsymbol{\Psi}(n-1), \boldsymbol{\Psi}(n) \in \mathbb{C}^{N\times K}$, such that

$$\operatorname{rank}\left\{\underbrace{\begin{bmatrix} \boldsymbol{\Psi}(n-\delta+1) & \cdots & \boldsymbol{\Psi}(n-1) & \boldsymbol{\Psi}(n) \end{bmatrix}}_{N\times\delta K}\right\} = \delta K, \tag{5.66}$$

where, as a consequence of the above matrix dimensions, we have

$$\delta K \leq N. \tag{5.67}$$

A necessary condition for $\mathbf{Z}^H\boldsymbol{\Theta}(n)$ be a full-row rank matrix is that $\widehat{\mathbf{Z}^H\boldsymbol{\Theta}}(n)$ be full row rank, which eventually requires that

$$\mu \leq \delta K \leq N, \tag{5.68}$$

since the rank of matrix $\widehat{\mathbf{Z}^H\boldsymbol{\Theta}}(n)$ is smaller than or equal to the ranks of each matrix that comprises the matrix product which defines $\widehat{\mathbf{Z}^H\boldsymbol{\Theta}}(n)$, and we already know that $(Q+B)K \geq L \geq \mu$.

Therefore, using

$$K \geq \left\lceil \frac{\mu}{\delta} \right\rceil, \tag{5.69}$$

with

$$\delta \leq \min\left\{Q+B, \left\lfloor \frac{N}{K} \right\rfloor\right\} \tag{5.70}$$

guarantees that $\widehat{\mathbf{Z}^H\boldsymbol{\Theta}}(n)$ is full row rank, which indicates that it would be possible to satisfy the original zero-forcing constraint as well.

The previous conditions are dependent on the congruous zeros of the channel. In order to develop a condition which is independent of the channel-zero locations, we could concentrate efforts on the worst case scenario, which occurs when $\mu = L$. Intuitively, this can be thought as the worst setup since the number of degrees of freedom in matrix $\mathbf{Z}^H\boldsymbol{\Theta}(n)$ is somehow minimized when $\mu = L$. This means that for such a case one has more chances of having a row of matrix $\mathbf{Z}^H\boldsymbol{\Theta}(n)$ described as a linear combination of its other rows. Similarly, to what we have performed before, we have

$$\begin{aligned}
\widetilde{\mathbf{Z}^H\boldsymbol{\Theta}}(n) &= \begin{bmatrix} \tilde{\mathbf{z}}_0^H\boldsymbol{\Psi}(n-Q-B+1) & \cdots & (c^*)^{Q+B-2}\tilde{\mathbf{z}}_0^H\boldsymbol{\Psi}(n-1) & (c^*)^{Q+B-1}\tilde{\mathbf{z}}_0^H\boldsymbol{\Psi}(n) \\ \vdots & \vdots & \cdots & \vdots \\ \tilde{\mathbf{z}}_{L-1}^H\boldsymbol{\Psi}(n-Q-B+1) & \cdots & (c^*)^{Q+B-2}\tilde{\mathbf{z}}_{L-1}^H\boldsymbol{\Psi}(n-1) & (c^*)^{Q+B-1}\tilde{\mathbf{z}}_{L-1}^H\boldsymbol{\Psi}(n) \end{bmatrix} \\
&= \underbrace{\begin{bmatrix} \tilde{\mathbf{z}}_0^H \\ \vdots \\ \tilde{\mathbf{z}}_{L-1}^H \end{bmatrix}}_{L\times N} \underbrace{\begin{bmatrix} \boldsymbol{\Psi}(n-Q-B+1) & \cdots & \boldsymbol{\Psi}(n-1) & \boldsymbol{\Psi}(n) \end{bmatrix}}_{N\times(Q+B)K} \\
&\quad \times \underbrace{\begin{bmatrix} \mathbf{I}_K & & & \\ & c^*\mathbf{I}_K & & \\ & & \ddots & \\ & & & (c^*)^{Q+B-1}\mathbf{I}_K \end{bmatrix}}_{(Q+B)K\times(Q+B)K} . \qquad (5.71)
\end{aligned}$$

Once again, by assuming that

$$L \leq \operatorname{rank}\left\{\begin{bmatrix} \boldsymbol{\Psi}(n-\delta+1) & \cdots & \boldsymbol{\Psi}(n-1) & \boldsymbol{\Psi}(n) \end{bmatrix}\right\} = \delta K \leq N, \qquad (5.72)$$

and by using $N = M + K \geq L$ (i.e., $B = 1$), or in other words,

$$M \geq L - K, \qquad (5.73)$$

we can achieve a ZF solution using $K = 1$ redundant elements, as long as (see Equation (5.64))

$$\min\{Q+1, N\} \geq \delta \geq L, \qquad (5.74)$$

which can be achieved with a large enough equalizer length ($Q \geq L - 1$), thus requiring many time-variant receiver matrices to estimate a block. Indeed, with such assumptions we can design a matrix $\widetilde{\mathbf{Z}^H\boldsymbol{\Theta}}(n)$ with rank L, which implies that we can design a matrix $\mathbf{Z}^H\boldsymbol{\Theta}(n)$ with rank L. This proves that the lower bound $K = 1$ on the number of transmitted redundant elements is achievable.

5.3.5 ROLE OF THE TIME-VARIANCE PROPERTY

It is important to point out the role of the time-variance property in the FIR reduced-redundancy systems. As an example, if we were working with LTI systems, then Equation (5.65) would have been written as follows:

$$\widehat{\mathbf{Z}^H\boldsymbol{\Theta}} = \begin{bmatrix} \tilde{\mathbf{z}}_0^H\boldsymbol{\Psi} & \cdots & (c^*)^{Q+B-2}\tilde{\mathbf{z}}_0^H\boldsymbol{\Psi} & (c^*)^{Q+B-1}\tilde{\mathbf{z}}_0^H\boldsymbol{\Psi} \\ \vdots & \vdots & \cdots & \vdots \\ \tilde{\mathbf{z}}_{\mu-1}^H\boldsymbol{\Psi} & \cdots & (c^*)^{Q+B-2}\tilde{\mathbf{z}}_{\mu-1}^H\boldsymbol{\Psi} & (c^*)^{Q+B-1}\tilde{\mathbf{z}}_{\mu-1}^H\boldsymbol{\Psi} \end{bmatrix}$$
$$= \underbrace{\begin{bmatrix} \tilde{\mathbf{z}}_0^H \\ \vdots \\ \tilde{\mathbf{z}}_{\mu-1}^H \end{bmatrix}}_{\mu\times N} \underbrace{\begin{bmatrix} \boldsymbol{\Psi} & \cdots & \boldsymbol{\Psi} & \boldsymbol{\Psi} \end{bmatrix}}_{N\times(Q+B)K} \underbrace{\begin{bmatrix} \mathbf{I}_K & & & \\ & c^*\mathbf{I}_K & & \\ & & \ddots & \\ & & & (c^*)^{Q+B-1}\mathbf{I}_K \end{bmatrix}}_{(Q+B)K\times(Q+B)K}. \tag{5.75}$$

Note that the rank of matrix $[\boldsymbol{\Psi} \ \cdots \ \boldsymbol{\Psi} \ \boldsymbol{\Psi}]$ is K. Hence, we would have to guarantee that

$$K \geq \mu, \tag{5.76}$$

in order to achieve a ZF solution, implying that $N = M + K \geq \mu$. This is a much stronger constraint than the previous one in expression (5.69) associated with LTV systems. For example, if all zeros of a channel transfer function are on the unitary circle, so that $\mu = L$, then we must use $K \geq L$ redundant elements in LTI systems, whereas we may still be able to use $K = 1$ in LTV systems!

5.4 TRANSCEIVERS WITH NO REDUNDANCY

As mentioned before, the ZF solution is suboptimal in the MSE sense. Indeed, an optimal MMSE-based solution could be achieved by jointly designing the transmitter and receiver filters. In this case, it would be possible to transmit without using any redundant elements at all. For example, a transceiver with memory and with *no redundancy* allows for the minimization of the MSE by jointly designing the transmitter and receiver FIR MIMO filters in the presence of, for example, near-end crosstalk and additive noise. This discussion is representative of other discussions related to some transceiver optimization, such as the ones described in [30, 83].

In the transceiver design we have to determine the matrices representing the transmitter and receiver. These matrices appear, for instance, in Equations (5.20) and (5.21), or in a more compact manner in Equation (5.22).

There are many ways to design the global transmitter and receiver matrices, respectively, denoted as $\mathcal{F}(n)$ (remember that this matrix has a block-Toeplitz structure as described in Equation (5.21)) and $\mathcal{G}(n)$. Here, we shall focus on minimizing the MSE of symbols, $\mathcal{E}_{\text{MSE}} \in \mathbb{R}_+$. The minimum MSE (MMSE) designs are very common in practical systems [83]. Consider that the transmitted vector $\overline{\mathbf{s}}(n)$ and the channel-noise vector $\overline{\mathbf{v}}(n)$ are, respectively, drawn from the zero-mean jointly wide-sense stationary (WSS) random sequences $\overline{\boldsymbol{s}}(n)$ and $\overline{\boldsymbol{v}}(n)$. Assuming that the estimated symbols are given by

$$\begin{aligned} \hat{\boldsymbol{s}}(n) &= \mathcal{G}(n)\left[\mathcal{HF}(n)\overline{\boldsymbol{s}}(n) + \overline{\boldsymbol{v}}(n)\right] \\ &= \mathcal{G}(n)\mathcal{HF}(n)\overline{\boldsymbol{s}}(n) + \mathcal{G}(n)\overline{\boldsymbol{v}}(n), \end{aligned} \tag{5.77}$$

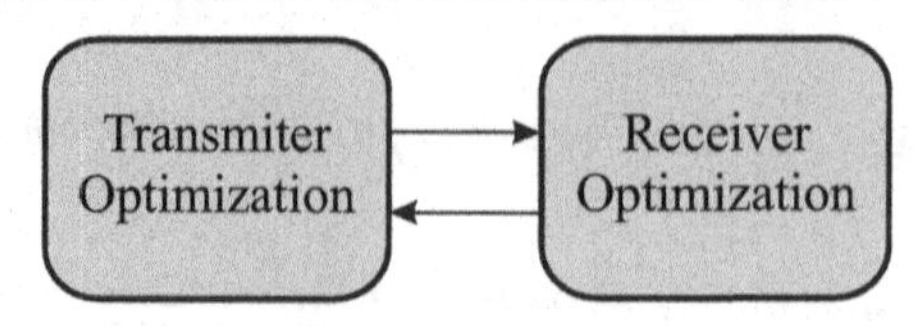

Figure 5.1: Optimization strategy.

then the overall MSE of symbols is given by (see also Equation (5.23))

$$\begin{aligned}\mathcal{E}_{\text{MSE}} &\triangleq E\{\|\hat{s}(n) - s(n)\|_2^2\} \\ &= E\{\|\mathcal{G}(n)\mathcal{H}\mathcal{F}(n)\overline{s}(n) + \mathcal{G}(n)\overline{v}(n) - \begin{bmatrix}\mathbf{0} & \mathbf{I}_M\end{bmatrix}\overline{s}(n)\|_2^2\} \\ &= \operatorname{tr}\left\{(\mathcal{G}(n)\mathcal{H}\mathcal{F}(n) - \begin{bmatrix}\mathbf{0} & \mathbf{I}_M\end{bmatrix})\mathbf{R}_{ss}(n)(\mathcal{G}(n)\mathcal{H}\mathcal{F}(n) - \begin{bmatrix}\mathbf{0} & \mathbf{I}_M\end{bmatrix})^H\right\} \\ &\quad + \operatorname{tr}\left\{\mathcal{G}(n)\mathbf{R}_{vv}(n)\mathcal{G}^H(n)\right\},\end{aligned} \tag{5.78}$$

in which $\mathbf{R}_{ss}(n) = E\{\overline{s}(n)\overline{s}^H(n)\}$ and $\mathbf{R}_{vv}(n) = E\{\overline{v}(n)\overline{v}^H(n)\}$. In addition, we have also assumed that $\overline{s}(n)$ and $\overline{v}(n)$ are uncorrelated, i.e., $\mathbf{R}_{sv} = E\{\overline{s}(n)\overline{v}^H(n)\} = E\{\overline{s}(n)\}E\{\overline{v}(n)\}^H = \mathbf{0}$.

In summary, it assumed that the FIR MIMO channel transfer matrices are known and that all the input signals to the system are uncorrelated with each other. In addition, we usually consider that the second order statistics of all the input signals are known. The transceiver is represented by FIR MIMO transfer matrices with given orders.

Our problem is to minimize MSE, usually subject to power constraints. The procedure to find a solution is to derive analytical expressions for the MSE criterion and power constraint as for example presented in [31]. For that we find the Karush-Kuhn-Tucker (KKT) conditions for optimality and implement an iterative numerical algorithm based on those necessary conditions.

The basic strategy is to design a transmitter assuming a known initial receiver and then design a receiver assuming the transmitter is the one previously designed, as Figure 5.1 illustrates. The procedure can be repeated until no further reduction in the MSE is achieved.

5.5 EXAMPLES

This section presents three examples that illustrate some of the concepts developed in this chapter. We will start with Example 5.4 in order to show a particular situation in which an FIR MIMO transceiver outperforms a CP-OFDM system, considering the bit-error rate (BER) as figure of merit. Then, Example 5.5 illustrates the BER performance of ZF- and MMSE-based systems satisfying or not the zero-forcing conditions described in Section 5.3. After that, we explain how the theoretical zero-forcing bounds derived here relate to the ZF detection of signals in long-code CDMA systems in Example 5.6.

Example 5.4 (OFDM & FIR MIMO Transceivers) This example assesses the BER *versus* signal-to-noise ratio (SNR) performance evaluated through a Monte Carlo simulation, where $M = 8$

BPSK symbols are transmitted per block through a channel whose transfer function is

$$H(z) = \left(1 - 0.9z^{-1}\right)\left(1 - 0.7e^{j2\pi 0.256}z^{-1}\right)\left(1 - 0.4e^{j2\pi 0.141}z^{-1}\right), \tag{5.79}$$

thus implying that $L = 3$. The number of redundant elements is always $K = L = 3$, which means that $N = 11$. The uncorrelated additive channel noise samples are complex, circularly symmetric, and Gaussian distributed. Two distinct systems are considered in this example: the CP-OFDM and the FIR MIMO transceiver proposed in [31], which optimizes the transmitter and receiver matrices through an iterative MMSE-based approach, such as the one briefly described in Section 5.4.

Figure 5.2 depicts the obtained results, in which it is clear the striking advantages of using FIR MIMO systems over OFDM-based systems in this particular setup. Of course, we are not taking into account implementation issues, such as computational complexity, modularity, just to mention a few. Indeed, a great advantage of OFDM is the low complexity implementation costs required to implement the fast Fourier and inverse fast Fourier transforms. □

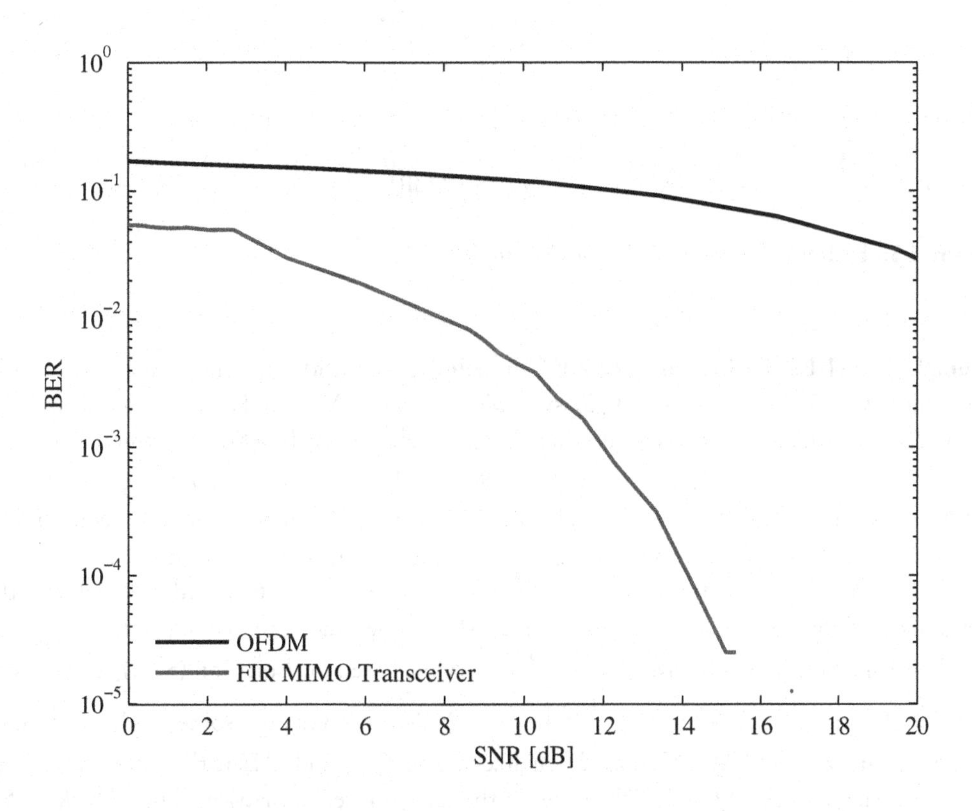

Figure 5.2: BER as a function of SNR for Example 5.4.

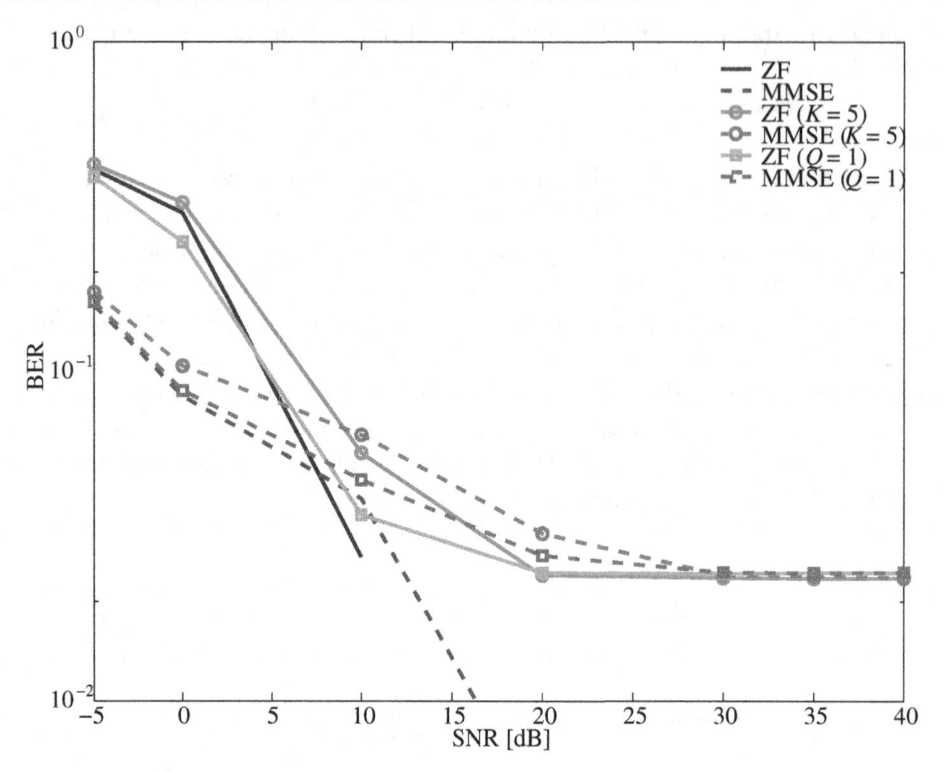

Figure 5.3: BER as a function of SNR for Example 5.5.

Example 5.5 (BER Performance & ZF Conditions) Consider the transmission of data blocks containing $N = 30$ entries, including M symbols of a 16-QAM digital constellation as well as $K = 30 - M$ redundant elements. Assume that the channel-transfer function has the following zeros: $1.8e^{j\frac{2\pi}{5}}$, $1.8e^{-j\frac{2\pi}{5}}$, $1.8e^{j\frac{3\pi}{5}}$, $1.8e^{-j\frac{3\pi}{5}}$, $1.8e^{j\frac{4\pi}{5}}$, $1.8e^{-j\frac{4\pi}{5}}$, $1.1e^{j\frac{7\pi}{8}}$, $1.1e^{-j\frac{7\pi}{8}}$, $0.9e^{j\frac{7\pi}{8}}$, $0.9e^{-j\frac{7\pi}{8}}$, $e^{j\frac{5\pi}{10}}$, $e^{-j\frac{5\pi}{10}}$, 0.8, which means that the channel order is $L = 13$, whereas the number of congruous zeros with respect to N is $\mu = 6$, corresponding to the first six zeros with magnitude 1.8.

In a time-invariant transmission, one must have $K \geq \mu = 6$ in order to achieve the ZF solution (see Subsection 5.3.5). This means that M must be smaller than or equal to 24. In addition, based on Equation (5.6), one has $B = \left\lceil \frac{13}{30} \right\rceil = 1$, while based on Equation (5.52), and considering that $T = 1$ and $K = 6$, one has $Q \geq \left\lceil \frac{13}{6} \right\rceil - 1 = 2$. Figure 5.3 depicts some results obtained using time-invariant ZF- and MMSE-based equalizers. The "ZF" and "MMSE" curves employ $K = 6$ redundant elements and $Q = 2$ as the order of the related receiver matrices. The "ZF ($K = 5$)" and "MMSE ($K = 5$)" curves employ $K = 5$ and $Q = 2$, which means that the number of redundant elements is violating the ZF constraint described in Section 5.3. The "ZF ($Q = 1$)" and "MMSE

($Q = 1$)" curves employ $K = 6$ and $Q = 1$, which means that the orders of the related receiver matrices are not satisfying the ZF constraint. The obtained results show that, by not following the ZF conditions, either by choosing $K = 5$ or $Q = 1$, a floor in the BER curves appears, starting from 20 dB of SNR. On the other hand, for the designs following the ZF conditions, the BER tends to zero as the SNR increases.

This example also illustrates that when alternative design criteria are used, such as the minimization of the MSE, the ZF conditions described in Section 5.3 are still useful in order to avoid performance loss due to errors in the reconstruction of the signal. This eventually means that, even for very high SNRs, it is still possible to have a "BER floor" due to the nonexistence of the ZF solution. On the other hand, this "BER floor" does not appear by observing the conditions presented in this chapter. □

Example 5.6 (Long-Code CDMA Systems) As previously illustrated in Figure 1.21 of Chapter 1, MIMO models encompass many block-transmission configurations, ranging from single-user point-to-point communications employing multiple antennas, to multiple-access schemes in multiuser systems. This example illustrates an application of the theoretical results related to FIR LTV transceivers with reduced redundancy within the framework of CDMA systems.

Indeed, CDMA systems can be described using the concept of MIMO transceivers. By using such a description it is possible to derive some theoretical conditions for designing equalizers that guarantee the perfect reconstruction of the transmitted signal at the receiver end. For example, we can apply the theoretical analysis in order to obtain multiuser detection in CDMA systems with long codes, i.e., codes which last for more than one symbol duration, as described in the following reasoning.

Consider the TMUX structure of Figure 2.6 of Chapter 2. Assume that $s_m(n)$ is the symbol associated with the mth user at the time instant n. It is possible to imagine that the impulse responses of the synthesis filters $f_m(k)$, with $m \in \mathcal{M}$, that appear in that figure can play the role of a *spreading sequence* of a CDMA system, whereas the impulse responses of analysis filters $g_m(k)$, with $m \in \mathcal{M}$, can be thought as the *de-spreading codes*, all of them associated with the mth user. If, in addition, we assume that the synthesis and analysis filters are actually time-varying filters, then we can consider that each subfilter implements a piece of a given CDMA spreading/de-spreading long code associated with a user. In other words, CDMA with long codes can be interpreted as CDMA with time-varying short codes.

As a result, it is possible to verify that CDMA with long codes can be represented by a time-varying structure, i.e., it is an example of FIR LTV system. Each signal block faces a time-varying code, which is implemented using time-varying transmit filters. Thus, by adapting the results concerning the amount of redundancy required to allow a ZF solution, it is possible to establish the conditions for the existence of zero-forcing multiuser detectors.

In this context, N represents the *spreading factor*, M denotes the number of codes that are going to be used, whereas K represents the number of unused codes. The main conclusion of this analysis is that ZF equalization is always possible in CDMA systems using long codes as long as the system is not at full capacity, i.e., one must have at least one unused code ($K \geq 1$). In addition, it can be shown that both the complexity of the receiver and its performance depend directly on the number of unused codes. The conditions derived serve as useful guidelines for the design of communications systems, allowing the tradeoff between performance and complexity of the receiver.

In conclusion, the existence of ZF equalizers is guaranteed if the amount of redundancy is greater than or equal to the number of congruous zeros or if there are enough different transmission filters at the transmitter. For practical channels, approximately congruous zeros may cause numerical instability in equalizer design. A CDMA system with long codes may be interpreted as a transmultiplexer with memoryless time-varying filters. In the uplink direction, the system can be modeled as a transmultiplexer with time-varying filters with memory. In any case, for the CDMA system, the block length is determined by the spreading factor, implying that the redundancy is equal to the number of unused codes. □

5.6 CONCLUDING REMARKS

This chapter addressed the problem of further reducing transmission redundancy in block transceivers by relieving their designs from the memoryless and time-invariance constraints. The use of time-varying transceivers with memory allows ZF solutions whose amount of required redundancy can be reduced to a single element. We discussed several results related to the amount of memory at the transmitter and receiver as well as the time variance of the transceiver in order to understand what are available to achieve ZF solutions with as little amount of redundancy in the transmission as possible. Despite this attractive feature some performance issues have to be considered particularly when the environment noise is not negligible. In this case, the MMSE solution is a practical solution where the ZF constraint can be relieved, leading to solutions where no redundancy is required. On the other hand, it is known that in many practical situations an increase in the amount of redundancy leads to reduction in the MSE. In summary, the actual designer challenge is to find the best trade-off between data throughput and system performance while keeping the transceiver implementation numerically robust and efficient.

Bibliography

[1] A. N. Akansu, P. Duhamel, X. Lin, and M. de Courville, "Orthogonal transmultiplexers in communication: a review," *IEEE Trans. on Signal Processing*, vol. 46, no. 4, pp. 979–995, April 1998. DOI: 10.1109/78.668551 Cited on page(s)

[2] A. N. Akansu and M. J. Medley, *Wavelets, Subband and Block Transforms in Communications and Multimedia*, 2nd Edition, Kluwer Academic Publishers, Norwell, MA, 1999. Cited on page(s)

[3] N. Al-Dhahir and S. N. Diggavi, "Guard sequence optimization for block transmission over linear frequency-selective channels," *IEEE Trans. Communications*, vol. 50, no. 6, pp. 938–946, June 2002. DOI: 10.1109/TCOMM.2002.1010613 Cited on page(s)

[4] J. B. Anderson, *Digital Transmission Engineering*, IEEE Press, New York, NY, 1999. Cited on page(s) 2

[5] A. Antoniou and W.-S. Lu, *Practical Optimization: Algorithms and Engineering Applications*. New York: Springer, 2008. Cited on page(s)

[6] L. Anttila, M. Valkama, and M. Renfors, "Circularity-based I/Q imbalance compensation in wideband direct- conversion receivers," *IEEE Trans. on Vehicular Technology*, vol. 57, no. 4, pp. 2099–2113, July 2008 DOI: 10.1109/TVT.2007.909269 Cited on page(s) 71

[7] J. R. Barry, E. A. Lee, and D. G. Messerschmitt, *Digital Communications*, 3rd edition, Kluwer Academic Publishers, Norwell, MA, 2004. Cited on page(s) 2

[8] T. Berger, *Rate Distortion Theory*, Prentice Hall, Englewood Cliffs, NJ, 1991. Cited on page(s)

[9] B. F.-Boroujeny, "OFDM versus filter bank multicarrier: development of broadband communication systems," *IEEE Signal Processing Magazine*, vol. 28, no. 3, pp. 92–112, May 2011. DOI: 10.1109/MSP.2011.940267 Cited on page(s)

[10] R. W. Chang, "High-speed multichannel data transmission with bandlimited orthogonal signals," *Bell. Sys. Tech. Journal*, vol. 45, pp. 1775–1796, December 1966. Cited on page(s) 55, 72

[11] R. W. Chang, "Orthogonal frequency multiplex data transmission systems," US-Patent, US3488445 (a), January 1970 Cited on page(s) 55, 72

[12] Y.-H. Chung and S.-M. Phoong, "Low complexity zero-padding zero-jamming DMT systems," in *Proc. of the 14th Eur. Signal Process. Conf. (EUSIPCO)*, Florence, Italy, September 2006, pp. 1–5. Cited on page(s)

[13] T. de Couasnon, R. Monnier, and J. B. Rault, "OFDM for digital TV broadcasting," *Signal Processing*, vol. 39, pp. 1–32, September 1994. DOI: 10.1016/0165-1684(94)90120-1 Cited on page(s) 55, 93, 94

[14] T. M. Cover and J. A. Thomas, *Elements of Information Theory*, 2nd edition, John Wiley, Hoboken, NJ, 2006. Cited on page(s) 2, 97

[15] Z. Cvetković, "Modulating waveforms for OFDM," in *Proc. of IEEE Int. Conf. Acoust., Speech, and Signal Process. (ICASSP)*, Phoenix, USA, March 1999, pp. 2463–2466. DOI: 10.1109/ICASSP.1999.760629 Cited on page(s)

[16] P. S. R. Diniz, *Adaptive Filtering: Algorithms and Practical Implementation*. 4th edition, Springer, New York, 2013. Cited on page(s)

[17] P. S. R. Diniz, E. A. B. da Silva, and S. L. Netto, *Digital Signal Processing: Systems Analysis and Design*, 2nd edition, Cambridge, UK, 2010. Cited on page(s) 30, 31, 34, 72, 141

[18] K-L. Du and M. N. Swamy, *Wireless Communication Systems: From RF Subsystems to 4G Enabling Technologies,* Cambridge University Press, Cambridge, UK, 2010. Cited on page(s)

[19] O. Edfors, M. Sandell, J.-J. van de Beek, D. Landström, and F. Sjöberg, "An introduction to orthogonal frequency-division multiplexing," *Internal Report Lulea University of Technology*, Sweden, September 1996. DOI: 10.1002/9780470031384.ch2 Cited on page(s)

[20] B. L. Floch, M. Alard, and C. Berrou, "Coded orthogonal frequency division multiplex," *Proceedings of IEEE*, vol. 83, no. 6, pp. 982–996, June 1995. DOI: 10.1109/5.387096 Cited on page(s) 93

[21] M. B. Furtado Jr., P. S. R. Diniz, and S. L. Netto, "Redundant paraunitary FIR transceivers for single-carrier transmission over frequency selective channels with colored noise," *IEEE Trans. on Communications*, vol. 55, no. 6, pp. 1125–1130, June 2007. DOI: 10.1109/TCOMM.2007.898826 Cited on page(s)

[22] R. G. Gallager, *Principles of Digital Communication*, Cambridge University Press, Cambridge, UK, 2008. Cited on page(s) 2, 95

[23] R. M. Gray, *Toeplitz and Circulant Matrix: A Review*, Now Publisher New York, NY, 2006. Cited on page(s)

[24] R. Guan and T. Stromer, "Krylov subspace algorithms and circulant-embedding method for efficient wideband single-carrier equalization," *IEEE Trans. on Signal Processing*, vol. 56, no. 6, pp. 2483–2495, June 2008. DOI: 10.1109/TSP.2007.912287 Cited on page(s)

[25] P. R. Halmos, *Finite-Dimensional Vector Spaces*. Boston: Springer, 1974. DOI: 10.1007/978-1-4612-6387-6 Cited on page(s)

[26] S. Haykin, *Communication Systems*, 4th Edition, John Wiley, New York, NY, 2001. Cited on page(s) 2, 95

[27] G. Heinig and K. Rost, "DFT representations of Toeplitz-plus-Hankel Bezoutians with application to fast matrix-vector multiplication," *Linear Algebra Appl.*, vol. 284, no. 1–3, pp. 157–175, November 1998. DOI: 10.1016/S0024-3795(98)10076-9 Cited on page(s) 123, 128, 132

[28] G. Heinig and K. Rost, "Hartley transform representations of inverses of real Toeplitz-plus-Hankel matrices," *Numer. Funct. Anal. and Optimiz.*, vol. 21, no. 1–2, pp. 175–189, February 2000. Cited on page(s) 123, 128

[29] G. Heinig and K. Rost, "Hartley transform representations of symmetric Toeplitz matrix inverses with application to fast matrix-vector multiplication," *SIAM J. Matrix Anal. Appl.*, vol. 22, no. 1, pp. 86–105, May 2000. DOI: 10.1137/S089547989833961X Cited on page(s)

[30] A. Hjørungnes, P. S. R. Diniz, and M. L. R. de Campos, "Jointly minimum BER FIR MIMO transmitter and receiver filters for binary signal vectors," *IEEE Trans. on Signal Processing*, vol. 52, no. 4, pp. 1021–1036, April 2004. DOI: 10.1109/TSP.2003.822291 Cited on page(s) 169

[31] A. Hjørungnes, M. L. R. de Campos, and P. S. R. and Diniz, "Jointly optimized transmitter and receiver filters FIR MIMO filters in the presence of near-end crosstalk," *IEEE Trans. on Signal Processing*, vol. 53, no. 1, pp. 346–359, January 2005. DOI: 10.1109/TSP.2004.838973 Cited on page(s) 170, 171

[32] A. Hjørungnes and P. S. R. Diniz, "Minimum BER FIR prefilter transform for communications systems with binary signaling and known FIR MIMO channel," *IEEE Signal Processing Letters*, vol. 12, no. 3, pp. 234–237, March 2005. DOI: 10.1109/LSP.2004.842272 Cited on page(s)

[33] M. Joham, *Optimization of Linear and Nonlinear Transmit Signal Processing*, Dr. Ing. Thesis, Technical University of Munich, Shaker Verlag, 2004. Cited on page(s)

[34] T. Kailath, S.-Y. Kung, and M. Morf, "Displacement ranks of a matrix," *Bulletin of The American Math. Soc.*, vol. 1, no. 5, pp. 769–773, September 1979. DOI: 10.1090/S0273-0979-1979-14659-7 Cited on page(s) 115

[35] T. Kailath and A. H. Sayed, "Displacement structure: theory and applications," *SIAM Review*, vol. 37, no. 3, pp. 297–386, September 1995. DOI: 10.1137/1037082 Cited on page(s) 115

[36] A. Lapidoth, *A Foundation in Digital Communication*, Cambridge, UK, 2009. Cited on page(s)

[37] M. Lazarus, "The great spectrum famine," *IEEE Spectrum*, vol. 47, no. 10, pp. 26–31, October 2010. DOI: 10.1109/MSPEC.2010.5583459 Cited on page(s) 149

[38] S. Lin and D. J. Costello, *Error Control Coding*, 2nd edition, Prentice Hall, Englewood Cliffs, NJ, 2004. Cited on page(s) 2, 93, 94

[39] Y.-P. Lin and S.-M. Phoong, "Perfect discrete multitone modulation with optimal transceivers," *IEEE Trans. on Signal Processing*, vol. 48, no. 6, pp. 1702–1711, June 2000. DOI: 10.1109/78.845928 Cited on page(s) 51

[40] Y.-P. Lin and S.-M. Phoong, "ISI-free FIR filter-bank transceivers for frequency selective channels," *IEEE Trans. on Signal Processing*, vol. 49, no. 11, pp. 2648–2658, November 2001. DOI: 10.1109/78.960412 Cited on page(s) 51

[41] Y.-P. Lin and S.-M. Phoong, "Optimal ISI-free DMT transceivers for distorted channels with colored noise," *IEEE Trans. on Signal Processing*, vol. 49, no. 11, pp. 2702–2712, November 2001. DOI: 10.1109/78.960417 Cited on page(s)

[42] Y.-P. Lin and S.-M. Phoong, "BER minimized OFDM systems with channel independent precoders," *IEEE Trans. on Signal Processing*, vol. 51, no. 9, pp. 2369–2380, September 2003. DOI: 10.1109/TSP.2003.815391 Cited on page(s)

[43] Y.-P. Lin and S.-M. Phoong, "OFDM transmitters: analog representation and DFT-based implementation," *IEEE Trans. on Signal Processing*, vol. 51, no. 9, pp. 2450–2453, September 2003. DOI: 10.1109/TSP.2003.815392 Cited on page(s) 55, 73

[44] Y.-P. Lin and S.-M. Phoong, "Minimum redundancy for ISI free FIR filter-bank transceivers," *IEEE Trans. on Signal Processing*, vol. 50, no. 4, pp. 842–853, April 2002. DOI: 10.1109/78.992130 Cited on page(s) 42, 51, 52, 107

[45] Y.-P. Lin, S.-M. Phoong, and P. P. Vaidyanathan, *Filter Bank Transceivers for OFDM and DMT Systems*, Cambridge University Press, Cambridge, UK, 2011. Cited on page(s) 42, 90, 95

[46] X. Ma, "Low-complexity block double-differential design for OFDM with carrier frequency offset," *IEEE Trans. on Communications* vol. 53, no. 12, pp. 2129–2138, December 2005. DOI: 10.1109/TCOMM.2005.860052 Cited on page(s) 70

[47] U. Madhow, *Fundamentals of Digital Communication*, Cambridge University Press, Cambridge, UK, 2008. Cited on page(s) 2

[48] W. A. Martins and P. S. R. Diniz, "Minimum redundancy multicarrier and single-carrier systems based on Hartley transform," in *Proc. of the 17th Eur. Signal Process. Conf. (EUSIPCO)*, Glasgow, Scotland, August 2009, pp. 661–665. Cited on page(s) 148

[49] W. A. Martins and P. S. R. Diniz, "Block-based transceivers with minimum redundancy," *IEEE Trans. on Signal Processing*, vol. 58, no. 3, pp. 1321–1333, March 2010. DOI: 10.1109/TSP.2009.2033000 Cited on page(s) 128

[50] W. A. Martins and P. S. R. Diniz, "Suboptimal linear MMSE equalizers with minimum redundancy," *IEEE Signal Processing Letters*, vol. 17, no. 4, pp. 387–390, April 2010. DOI: 10.1109/LSP.2010.2042515 Cited on page(s) 148

[51] W. A. Martins and P. S. R. Diniz, "Pilot-aided designs of memoryless block equalizers with minimum redundancy," in *Proc. of the IEEE Int. Symp. Circuit. Syst. (ISCAS)*, Paris, France, May 2010, pp. 275–279. DOI: 10.1109/ISCAS.2010.5537975 Cited on page(s) 142

[52] W. A. Martins and P. S. R. Diniz, "Memoryless block transceivers with minimum redundancy based on Hartley transform," *Signal Processing*, vol. 91, no. 2, pp. 240–251, February 2011. DOI: 10.1016/j.sigpro.2010.06.026 Cited on page(s) 148

[53] W. A. Martins and P. S. R. Diniz, "Analysis of Zero-Padded Optimal Transceivers," *IEEE Trans. on Signal Processing*, vol. 59, no. 11, pp. 5443–5457, November 2011. DOI: 10.1109/TSP.2011.2162327 Cited on page(s)

[54] W. A. Martins and P. S. R. Diniz, "LTI transceivers with reduced redundancy," *IEEE Trans. on Signal Processing*, vol. 60, no. 2, pp. 766–780, February 2012. DOI: 10.1109/TSP.2011.2174056 Cited on page(s) 142

[55] B. Muquet, Z. Wang, G. B. Giannakis, M. de Courville, and P. Duhamel, "Cyclic prefixing or zero padding wireless multicarrier transmissions?," *IEEE Trans. on Communications*, vol. 50, no. 12, pp. 2136–2148, December 2002. DOI: 10.1109/TCOMM.2002.806518 Cited on page(s) 50, 90, 91, 141, 143

[56] A. F. Naguib, *Adaptive Antennas for CDMA Wireless Networks*, Ph.D. Thesis, Stanford University, Stanford, CA, August 1996. Cited on page(s)

[57] H. H. Nguyen and E. Shwedyk, *A First Course in Digital Communications*, Cambridge University Press, Cambridge, UK, 2009. Cited on page(s) 2

[58] S. Ohno, "Performance of single-carrier block transmissions over multipath fading channels with linear equalization," *IEEE Trans. on Signal Processing*, vol. 54, no. 10, pp. 3678–3687, October 2006. DOI: 10.1109/TSP.2006.879321 Cited on page(s)

[59] S. Ohno and G. B. Giannakis, "Optimal training and redundant precoding for block transmissions with application to wireless OFDM," *IEEE Trans. Communications*, vol. 50, no. 12, pp. 2113–2123, December 2002. DOI: 10.1109/TCOMM.2002.806547 Cited on page(s)

[60] V. Y. Pan, "A unified superfast divide-and-conquer algorithm for structured matrices over abstract fields," *MSRI Preprint 1999-033, Mathematical Science Research Institute*, Berkeley, CA, 1999. Cited on page(s)

[61] V. Y. Pan, *Structured Matrices and Polynomials: Unified Superfast Algorithms*, Springer, New York, NY, 2001. DOI: 10.1007/978-1-4612-0129-8 Cited on page(s) 115, 117

[62] V. Y. Pan and X. Wang, "Inversion of displacement operators," *SIAM J. Matrix Anal. Appl.*, vol. 24, no. 3, pp. 660–667, 2003. DOI: 10.1137/S089547980238627X Cited on page(s)

[63] F. Pancaldi, G. Vitetta, R. Kalbasi, N. Al-Dhahir, M. Uysal, and H. Mheidat, "Single-carrier frequency domain equalization," *IEEE Signal Processing Magazine*, vol. 25, no. 5, pp. 37–56, September 2008. DOI: 10.1109/MSP.2008.926657 Cited on page(s) 30

[64] A. Peled and A. Ruiz, "Frequency domain data transmission using reduced computational complexity algorithms," in *Proc. of IEEE Int. Conf. Acoust., Speech, and Signal Process. (ICASSP)*, Denver, USA, April 1980, pp. 964–967. DOI: 10.1109/ICASSP.1980.1171076 Cited on page(s) 55, 72

[65] J. G. Proakis, *Digital Communications*, 4th edition, McGraw-Hill, New York, NY, 2001. Cited on page(s) 2

[66] J. G. Proakis and M. Salehi, *Communication Systems Engineering*, Prentice Hall, Upper Saddle River, NJ, 1994. Cited on page(s) 2, 95

[67] T. S. Rappaport, *Wireless Communications: Principles and Practice*, 2nd Edition, Prentice Hall, Upper Saddle River, NJ, 2002. Cited on page(s) 17

[68] C. B. Ribeiro, M. L. R. de Campos, and P. S. R. Diniz, "FIR equalizers with minimum redundancy," in *Proc. of IEEE Int. Conf. Acoust., Speech, and Signal Process. (ICASSP)*, Orlando, USA, May 2002, pp. 2673–2676. DOI: 10.1109/ICASSP.2002.5745198 Cited on page(s)

[69] C. B. Ribeiro, M. L. R. de Campos, and P. S. R. Diniz, "Zero-forcing multiuser detection in CDMA systems using long codes," in *Proc. of IEEE Global Telecom. Conf. (Globecom)*, San Francisco, USA, December 2003, pp. 2463–2467. DOI: 10.1109/GLOCOM.2003.1258679 Cited on page(s)

[70] C. B. Ribeiro, M. L. R. de Campos, P. S. R. Diniz, "Time-varying FIR transmultiplexers with minimum redundancy," *IEEE Trans. on Signal Processing*, vol. 57, no. 3, pp. 1113–1127, March 2009. DOI: 10.1109/TSP.2008.2010007 Cited on page(s) 36, 156

[71] A. Scaglione, S. Barbarossa, and G. B. Giannakis, "Filter-bank transceivers optimizing information rate in block transmissions over dispersive channels," *IEEE Trans. on Information Theory*, vol. 45, no. 3, pp. 1019–1032, April 1999. DOI: 10.1109/18.761338 Cited on page(s)

[72] A. Scaglione, G. B. Giannakis and S. Barbarossa, "Redundant filter-bank precoders and equalizers part I: unification and optimal designs," *IEEE Trans. on Signal Processing*, vol. 47, no. 7, pp. 1988–2006, July 1999. DOI: 10.1109/78.771047 Cited on page(s) 36

[73] A. Scaglione, G. B. Giannakis and S. Barbarossa, "Redundant filterbank precoders and equalizers part II: blind channel equalization, synchronization, and direct equalization," *IEEE Trans.*

on Signal Processing, vol. 47, no. 7, pp. 2007–2022, July 1999. DOI: 10.1109/78.771048 Cited on page(s)

[74] A. Schoonen, *IQ imbalance in OFDM Wireless LAN systems*, M.Sc. Thesis, Eindhoven University of Technology, Netherlands, 2006. Cited on page(s) 71

[75] B. Sklar, *Digital Communications: Fundamentals and Applications*, 2nd edition, Prentice Hall, Upper Saddle River, NJ, 2001. Cited on page(s) 2

[76] T. Starr, J. M. Cioffi, and P. J. Silverman, *Understanding Digital Subscriber Line Technology*, Prentice Hall, Upper Saddle River, NJ, 1999. Cited on page(s)

[77] G. Strang, *Linear Algebra and Its Applications*, Harcourt Brace Jovanovich San Diego, CA, 1988. Cited on page(s)

[78] A. Tarighat, R. Bagheri, and A. H. Sayed, "Compensation schemes and performance analysis of IQ imbalances in OFDM receivers ," *IEEE Trans. on Signal Processing*, vol. 53, no. 8, pp. 3257–3268, August 2005. DOI: 10.1109/TSP.2005.851156 Cited on page(s) 71

[79] S. Trautmann and N. J. Fliege, "Perfect equalization for DMT systems without guard interval," *IEEE J. Sel. Areas in Communications*, vol. 20, no. 5, pp. 987–996, June 2002. DOI: 10.1109/JSAC.2002.1007380 Cited on page(s)

[80] D. Tse and P. Viswanath, *Fundamentals of Wireless Communications*, Cambridge University Press, Cambridge, UK, 2005. Cited on page(s) 17, 107

[81] P. P. Vaidyanathan, *Multirate Systems and Filter Banks*, Prentice Hall, Englewood Cliffs, NJ, 1993. Cited on page(s) 34

[82] P. P. Vaidyanathan, "Filter banks in digital communications," *IEEE Circuits and Systems Magazine*, vol. 1, pp. 4–25, 2001. DOI: 10.1109/MCAS.2001.939098 Cited on page(s)

[83] P. P. Vaidyanathan, S.-M. Phoong, and Y.-P. Lin, *Signal Processing and Optimization for Transceiver Systems*, Cambridge University Press, Cambridge, UK, 2010. DOI: 10.1017/CBO9781139042741 Cited on page(s) 49, 90, 95, 126, 144, 169

[84] P. P. Vaidyanathan and B. Vrcelj, "Transmultiplexers as precoders in modern digital communication: a tutorial review," in *Proc. of the IEEE Int. Symp. Circuit. Syst. (ISCAS)*, Vancouver, Canada, May 2004, pp. V-405–V-412. DOI: 10.1109/ISCAS.2004.1329590 Cited on page(s)

[85] M. Valkama, M. Renfors, and V. Koivunen, "Advanced methods for I/Q imbalance compensation in communication receivers," *IEEE Trans. on Signal Processing*, vol. 49, no. 10, pp. 2335–2344, October 2001. DOI: 10.1109/78.950789 Cited on page(s) 71

[86] Z. Wang, and G. B. Giannakis, "Wireless multicarrier communications," *IEEE Signal Processing Magazine*, vol. 17, no. 3, pp. 29–48, May 2000. DOI: 10.1109/79.841722 Cited on page(s)

[87] Z. Wang, X. Ma, and G. B. Giannakis, "OFDM or single-carrier block transmissions?," *IEEE Trans. on Communications*, vol. 52, no. 3, pp. 380–394, March 2004. DOI: 10.1109/TCOMM.2004.823586 Cited on page(s) 30, 91

[88] S. B. Weinstein, "The history of orthogonal frequency-division multiplexing," *IEEE Communications Magazine*, vol. 47, no. 11, pp. 26–35, November 2009. DOI: 10.1109/MCOM.2009.5307460 Cited on page(s) 55

[89] S. B. Weinstein and P. M. Ebert, "Data transmission by frequency-division multiplexing using the discrete Fourier transform," *IEEE Trans. on Communication Technology*, vol. 19, no. 5, pp. 628–634, October 1971. DOI: 10.1109/TCOM.1971.1090705 Cited on page(s) 55, 72

[90] S. B. Wicker, *Error Control Systems for Digital Communication and Storage*, Prentice Hall, Upper Saddle River, NJ, 1994. Cited on page(s) 2, 93

[91] X. G. Xia, "New precoding for intersymbol interference cancellation using nonmaximally decimated multirate filterbanks with ideal FIR equalizers," *IEEE Trans. on Signal Processing*, vol. 45, no. 10, pp. 2431–2441, October 1997. DOI: 10.1109/78.640709 Cited on page(s)

[92] Y. Yao and G. B. Giannakis, "Blind carrier frequency offset estimation in SISO, MIMO, and multiuser OFDM systems," *IEEE Trans. on Communications* vol. 53, no. 1, pp. 173–183, January 2005. DOI: 10.1109/TCOMM.2004.840623 Cited on page(s) 70

[93] W. Zhang, X. Ma, B. Gestner, and D. V. Anderson, "Designing low-complexity equalizers for wireless systems," *IEEE Communications Magazine*, vol. 47, no. 1, pp. 56–62, January 2009. DOI: 10.1109/MCOM.2009.4752677 Cited on page(s)

[94] *Evolved Universal Terrestrial Radio Access (E-UTRAN): Multiplexing and Channel Coding*, 3GPP TS 36212, ver. 8.7.0, 3rd Generation Partnership Project, May 2009. Cited on page(s) 144

Authors' Biographies

PAULO S. R. DINIZ

Paulo S. R. Diniz was born in Niterói, Brazil. He received the Electronics Eng. degree (Cum Laude) from the Federal University of Rio de Janeiro (UFRJ) in 1978, a M.Sc. degree from COPPE/UFRJ in 1981, and a Ph.D. from Concordia University, Montreal, P.Q., Canada, in 1984, all in Electrical Engineering.

Since 1979 he has been with the Department of Electronic Engineering (undergraduate) at UFRJ. He has also been with the Program of Electrical Engineering (the graduate studies dept.), COPPE/UFRJ, since 1984, where he is presently a Professor. He served as Undergraduate Course Coordinator and Chairman of the Graduate Department. He has received the Rio de Janeiro State Scientist award from the Governor of Rio de Janeiro.

From January 1991 to July 1992, he was a visiting Research Associate in the Department of Electrical and Computer Engineering of University of Victoria, Victoria, B.C., Canada. He also held a Docent position at Helsinki University of Technology. From January 2002 to June 2002, he was a Melchor Chair Professor in the Department of Electrical Engineering of University of Notre Dame, Notre Dame, IN, USA. His teaching and research interests are in analog and digital signal processing, adaptive signal processing, digital communications, wireless communications, multirate systems, stochastic processes, and electronic circuits. He has published several refereed papers in some of these areas and wrote the books *Adaptive Filtering: Algorithms and Practical Implementation*, 4th ed., Springer, NY, 2012, and *Digital Signal Processing: System Analysis and Design*, 2nd ed., Cambridge University Press, Cambridge, UK, 2010 (with E.A.B. da Silva and S.L. Netto).

He has served as the Technical Program Chair of the 1995 MWSCAS held in Rio de Janeiro, Brazil. He was the General co-Chair of the IEEE ISCAS2011, and Technical Program co-Chair of the IEEE SPAWC2008. He has been on the technical committee of several international conferences including ISCAS, ICECS, EUSIPCO, and MWSCAS. He has served as Vice President for region 9 of the IEEE Circuits and Systems Society and as Chairman of the DSP technical committee of the same Society. He is also a Fellow of IEEE. He has served as associate editor for the following Journals: *IEEE Transactions on Circuits and Systems II: Analog and Digital Signal Processing* from 1996–1999, *IEEE Transactions on Signal Processing* from 1999–2002, and the *Circuits, Systems and Signal Processing Journal* from 1998–2002. He was a distinguished lecturer of the IEEE Circuits and Systems Society from 2000–2001. In 2004, he served as distinguished lecturer of the IEEE Signal

Processing Society and received the 2004 Education Award of the IEEE Circuits and Systems Society. He also holds some best-paper awards from conferences and from an IEEE journal.

WALLACE A. MARTINS

Wallace A. Martins was born in Brazil in 1983. He received an Electronics Engineering degree (Cum Laude) from the Federal University of Rio de Janeiro (UFRJ) in 2007, and M.Sc. and D.Sc. degrees in Electrical Engineering from COPPE/UFRJ in 2009 and 2011, respectively. He worked as a technical consultant for Nokia Institute of Technology (INDT), Brazil, and for TechKnowledge Training, Brazil. In 2008, he was a research visitor at the Department of Electrical Engineering, University of Notre Dame, Notre Dame, IN. Since 2010 he has been with the Department of Control and Automation Industrial Engineering, Federal Center for Technological Education Celso Suckow da Fonseca (CEFET/RJ – UnED-NI), where he is presently a Lecturer of Engineering. His research interests are in the fields of digital communication, microphone array signal processing, and adaptive signal processing. Dr. Martins received the Best Student Paper Award from EURASIP at EUSIPCO-2009, Glasgow, Scotland.

MARKUS V. S. LIMA

Markus V. S. Lima was born in Rio de Janeiro, Brazil in 1984. He received an Electronics Engineering degree from the Federal University of Rio de Janeiro (UFRJ) in 2008, an M.Sc. degree in Electrical Engineering from COPPE/UFRJ in 2009, and is currently pursuing his D.Sc. degree at COPPE/UFRJ. He has served as a teaching assistant for the following undergraduate courses taught at UFRJ: Digital Transmission, Digital Signal Processing, and Linear Systems. His main interests are in adaptive signal processing, microphone array signal processing, digital communications, wireless communications, statistical signal processing, and linear algebra.